Synthese Library

Studies in Epistemology, Logic, Methodology, and Philosophy of Science

Volume 395

The aim of *Synthese Library* is to provide a forum for the best current work in the methodology and philosophy of science and in epistemology. A wide variety of different approaches have traditionally been represented in the Library, and every effort is made to maintain this variety, not for its own sake, but because we believe that there are many fruitful and illuminating approaches to the philosophy of science and related disciplines.

Special attention is paid to methodological studies which illustrate the interplay of empirical and philosophical viewpoints and to contributions to the formal (logical, set-theoretical, mathematical, information-theoretical, decision-theoretical, etc.) methodology of empirical sciences. Likewise, the applications of logical methods to epistemology as well as philosophically and methodologically relevant studies in logic are strongly encouraged. The emphasis on logic will be tempered by interest in the psychological, historical, and sociological aspects of science.

Besides monographs *Synthese Library* publishes thematically unified anthologies and edited volumes with a well-defined topical focus inside the aim and scope of the book series. The contributions in the volumes are expected to be focused and structurally organized in accordance with the central theme(s), and should be tied together by an extensive editorial introduction or set of introductions if the volume is divided into parts. An extensive bibliography and index are mandatory.

More information about this series at http://www.springer.com/series/6607

Fabrice Correia • Sven Rosenkranz

Nothing To Come

A Defence of the Growing Block Theory of Time

Springer

Fabrice Correia
Département de Philosophie
Université de Genève
Genève, Switzerland

Sven Rosenkranz
ICREA
Barcelona, Spain

Departament de Filosofia
Universitat de Barcelona
Barcelona, Spain

Synthese Library
ISBN 978-3-030-08768-5   ISBN 978-3-319-78704-6 (eBook)
https://doi.org/10.1007/978-3-319-78704-6

Printed on acid-free paper

This Springer imprint is published by the registered company Springer International Publishing AG part of Springer Nature.
The registered company address is: Gewerbestrasse 11, 6330 Cham, Switzerland

Introduction

This monograph undertakes to give a limited defence of the Growing Block Theory of time (GBT), as first conceived by C. D. Broad in his 1923 book *Scientific Thought*. The defence is limited in that we do not aim to show that GBT is better than its rivals. Rather, we merely intend to show that GBT is better than proponents of rival views make it out to be. We set out to do so by showing that there is a coherent, logically perspicuous and ideologically lean formulation of GBT that is suited to successfully answer certain philosophically motivated arguments against it that tend to dominate the literature on the growing block.

As a first rough approximation, GBT is the view according to which what there is increases as time goes by, with new things being added along the way, all the while nothing is lost in the process. It is likewise part of this view that new additions to what there is are located at the time of their addition, and that there never is any time succeeding what is new in this sense. There thus always is an edge of becoming beyond which there is literally nothing in time which is as yet to come.

Although Broad's original text is refreshingly clear, GBT has received a rather bad press ever since its inception. To a large extent this is owing to a failure to read Broad (1923) very closely and with the required dose of charity. Thus, GBT has variably been charged with being an ill-conceived hybrid between presentism about the future and eternalism about the past, with commitment to past things having an unacceptable, because utterly mysterious, 'zombie'-status, and above all with inviting an unpalatable scepticism about the location of the edge of becoming that it posits. None of these charges can ultimately be substantiated – or so we shall argue.

Other objections derive from the commentators' tendency to exclusively focus on later, in our opinion misconceived attempts to articulate the view, at the cost of entirely ignoring Broad's original contribution – for instance, the charge that GBT allows for hostile takeover by eternalists or the charge that GBT requires two time-dimensions, or two notions of tense, or some kind of novel predication that is neither tensed nor tenseless. We argue that, properly understood, GBT invites none of these charges. As regards the former, we show that GBT incurs commitment to claims that no eternalist is willing to make. As regards the latter, we show that GBT has no need for further time-dimensions, a duplication of tenses, or any other exotic machinery unfamiliar from other quarters. In fact, GBT does not even have any need for the notions of being present, past or future, which should be welcomed by anyone who, like Timothy Williamson, considers these notions as too obscure to cut any metaphysical ice.

GBT offers a dynamic view of temporal reality according to which what facts there are varies with time. The facts it takes to vary across time are chiefly facts about what exists. This is a feature GBT shares with presentism – the other of the two most prominent tensed ontologies. Both are varieties of what Williamson calls *temporaryism*, i.e. the view according to which sometimes some things sometimes do not exist. By contrast, views such as the Moving Spotlight Theory combine their dynamic conception of temporal reality with an unchanging ontology. Like eternalism, which promotes a thoroughly static conception of temporal reality, such a view implies what Williamson calls *permanentism*, i.e. the doctrine that always everything always exists. Both presentism and the Moving Spotlight Theory of time would seem to heavily rely on the notion of being present. This, too, turns out to be an unfounded prejudice, or so we shall argue. There are after all formulations of either type of view that, just like the version of GBT we favour, have no need for presentness.

GBT is often said to be motivated by the thought that while the future is open the past is fixed and that by accepting presently existing things last located in the past, one has an easier time to account for truths about the past than one would otherwise have. If this were so, it would suggest that GBT can claim a clear advantage over presentism that shuns such past objects. Similarly, however, it would then seem that proponents of GBT have an equally hard time to account for contingent truths about the future, if such there be. Of course, one radical view of the open future holds that future contingents are neither true nor false, in which case this would prove no drawback; and Broad (1923) himself argued for such a failure of bivalence for future contingents. By contrast, we shall argue that this sort of diagnosis rests on too strong a conception of the grounding requirement on tensed truths, and that once this is taken to heart and a more sensible requirement is tabled, presentism and GBT alike can be shown to be equally well positioned to account for truths about the past. Similarly, both views can accommodate bivalence for future contingents and allow for a radical sense in which the future is open – a sense unavailable to their permanentist opponents. However, only GBT can avail itself of this sense of openness while retaining the asymmetry between the future as open and the past as fixed: for the presentist, if the future is said to be open in this sense, so must be the past.

Relativistic physics, with its rejection of absolute simultaneity among spatially distant events, poses a threat to classical theories of time. It often goes unnoticed that the challenge not only afflicts dynamic views, but also static views, of which the so-called B-theory of time serves as the primary example. Several authors have tried to show that dynamic views, and presentism in particular, can be made to cohere with the insights of relativistic physics. We argue that prominent such attempts prove ultimately unsuccessful.[1] Admittedly, this is not conclusive evidence that classical theories of time are doomed to failure – to the extent, namely, that it is so far unclear whether physics itself may evolve to rehabilitate a privileged foliation of spacetime that reinstates an absolute and total temporal order. However, as we shall argue, one's metaphysical views should not be hostage to such empirical fortune; instead, one should prepare for the worst case, i.e. the case in which there simply is no physically respectable sense at all in which two distant events may be said to be absolutely simultaneous. Accordingly, we suggest that classical theories should be subjected to a systematic overhaul, leading to relativistic versions of them that retain as much as possible of the spirit of their predecessors, yet forego any commitment to an absolute and total temporal order. The finding that classical theories can be articulated using only quantification, identity, temporal operators and relations, to be substantiated in what follows, already suggests a way forward in the project of implementing this revisionary strategy: temporal relations should systematically be replaced by spatiotemporal relations, temporal operators by spacetime operators, and existence claims should be understood to be spacetime- rather than time-sensitive. Proceeding from this set of ideas, we devise formulations of relativistic counterparts to GBT, presentism and the other contenders, and discuss their relative merits and shortcomings.

The book is most naturally seen to divide into three main parts. The first part, which comprises Chaps. 1, 2 and 3, provides the background logic we will use throughout, and elucidates basic notions that will play a crucial role in the remainder of the book. The middle part consists of Chaps. 4 and 5. In Chap. 4, we reconstruct and critically discuss C. D. Broad's original version of GBT, before we then introduce our own version and set it apart from more recent attempts to articulate the view. In Chap. 5, we devise formulations of GBT's rivals, using the same logico-conceptual framework. The last part, which comprises Chaps. 6, 7, 8 and 9, addresses three distinct challenges that have been marshalled against GBT in particular or dynamic views in general. In three technical appendices, we provide semantic characterisations of classical theories of time, and of their relativistic successor theories, and show how classical theories can be derived from their relativistic counterparts on assumption of principles doomed to fail in relativistic spacetime.

To be more specific, in Chap. 1, we first introduce the operator approach to tense according to which tenses, simple or complex, are represented by operators or combinations thereof; secondly, we devise the propositional fragment of a tense-logic which is, except for its commitment to the linearity of time and the transitivity of the relation of precedence, minimal; and thirdly, we precisify what it means to take tense seriously for the purposes of metaphysical enquiry.

Chapter 2 then introduces the distinction between permanentism and temporaryism as two broad classes of competing views on existence and time, presents a

[1] We will not discuss Kit Fine's fragmentalism, however (Fine 2005). Given its commitment to reality's being irremediably incoherent, it is unclear to what extent opting for fragmentalism is any less revisionary than pursuit of the – coherence-abiding – strategy we shall ultimately recommend.

combined theory of quantification and identity that is as yet neutral between these types of views, and argues that either type of view perfectly coheres with the idea that quantification is absolutely unrestricted.

In Chap. 3, we explicate two relations, that of temporal location and that of precedence and devise a number of postulates governing either relation. This concludes the first, preliminary part of the book.

In Chap. 4, we set out Broad's version of GBT which, though mostly on the right track, proves to have certain shortcomings that lead us to devise an improved formulation of the view. After proving a number of important theorems, we then briefly discuss the alternative versions of GBT respectively advanced by Michael Tooley and Tim Button, and argue that our own version of GBT is superior.

In Chap. 5, we critically review extant formulations of presentism and the Moving Spotlight Theory, and having found them wanting, offer what we consider to be better versions of either view that still remain faithful to the spirit of the original proposals.

In Chap. 6, we critically examine and defuse the notorious *epistemic objection*, foreshadowed by David Lewis, first properly formulated by Craig Bourne, and further fleshed out by David Braddon-Mitchell and Trenton Merricks. We argue that this objection misfires on several counts and show how, based on our knowledge of GBT, we can now know that now is on the edge of becoming.

Chapter 7 addresses another slightly more involved challenge, the *semantic objection*, that derives from the combination of four elements: the idea that there are future contingents, the charge that owing to the grounding requirement on truth, proponents of GBT are bound to consider such future contingents as being neither true nor false, the thought that supervaluationism gives the best account of the failure of bivalence for future contingents, and lastly the claim that GBT is incompatible with supervaluationism. We reject the second element and argue that it rests on an unreasonably strong version of the grounding requirement. Replacing the latter by a weaker and more plausible requirement on tensed truths, we argue that GBT is perfectly consistent with the unrestricted validity of bivalence. Its endorsement of bivalence notwithstanding, GBT nonetheless allows for a sense in which the future might be open that no permanentist will be willing to grant; and while presentists may likewise invoke this conception of openness, they are – unlike proponents of GBT – committed to treating the past as equally open.

In Chap. 8, we articulate the challenge, posed by relativistic physics, according to which there is no sense, consonant with physical theory, in which there may be said to be an absolute and total temporal order. We show that this challenge targets all classical views on time, including standard eternalism, and that appeal to temporal operators is no help in the attempt to dodge it. We proceed to critically examine a recent proposal by Dean Zimmerman which seeks to find room for a privileged foliation of spacetime into spacelike hypersurfaces in the contingent distribution of occupants of spacetime rather than the structure of spacetime itself. According to our diagnosis, this attempt likewise fails.

In Chap. 9, at last, we take a radical turn and propose to revise classical theories of time with the aim of arriving at relativistically acceptable successor theories of the main contenders. To this end, we devise a spacetime-sensitive language and a spatiotemporal logic that presupposes no more than the fourfold causal structure familiar from both Special and General Relativity. With these preliminaries in place, we then motivate the family of views we subsume under the label of *spatiotemporaryism*, discerning a certain continuity with the rationale of their prerelativistic precursors, and offer particular varieties of spatiotemporaryism that mimic as closely as possible the classical theories that they are intended to replace. It turns out that while there are two versions each of relativistic presentism and relativistic GBT, only one of each pair can rightly be said to serve as a relativistic counterpart to the classical view it respectively supersedes. Given the conjunction of two principles, invalid in relativistic spacetime, as further premises, as well as a suitable translation function, the relativistic counterparts can be shown to imply their classical precursors.

We first started to systematically explore GBT in the context of discussions about the open future – an issue we had left aside in our first joint book on A-theories of time (Correia and Rosenkranz 2011). It was our dissatisfaction with this omission that originally led us to think about a conception of the open future that might rightly be labelled 'the doomsday conception'. By that time, we believed to have a good argument for discarding GBT as incoherent, and so sought to combine the doomsday conception with a variant of the Moving Spotlight Theory of time. However, our confidence in the tenability of that argument soon crumbled. But it was only through detailed study of Broad's own work and familiarisation with the somewhat disappointing state of philosophical debate about the growing block, that we came to appreciate that, within the A-theorists' camp at least, GBT is a real contender.

These efforts resulted in a first article on the growing block that defended it against the attacks by Bourne, Braddon-Mitchell and Merricks (Correia and Rosenkranz 2013). This article prompted a response by Braddon-Mitchell to which we were given the opportunity to reply (Braddon-Mitchell 2013; Correia and Rosenkranz 2015a). Braddon-Mitchell's response made us even more determined that it was time for a systematic study of the view at book-length.

Reading Broad's work on time is stimulating and instructive despite its age – and sometimes also revealing, which is why we highly recommend it. Not only did he anticipate – almost *verbatim* – Prior's 'Thank goodness that's over' argument (Broad 1938: 267, cf. also 527-33; Prior 1959). He likewise anticipated the alleged problem of the rate of passage that Prior (1958) conjectured was first articulated by Smart (1949) (Broad 1938: 277), as well as the suggestion that higher time dimensions might be needed in order to make room for time's passage – a problem whose conception is commonly attributed to Smart (1949) and D. C. Williams (1951a) (Broad 1938: 277-280).

We are confident that Prior (1959) put the latter two problems successfully to rest, while we here nowhere rely on the argument from relief, if only because we do not even aim to show that dynamic views are superior to static views. It has proved already hard enough to devise perspicuous formulations of the different contenders and a logico-conceptual framework in which to frame the discussion, to defuse arguments against GBT and other dynamic views by showing them to rest on unwarranted or misguided assumptions, and to offer relativistically acceptable variants of classical views whose systematic articulation promises to further the metaphysical debate.

We would like to express our heartfelt thanks to those colleagues and friends whose critical feedback has helped to shape our thoughts on the subject over the years. Thus we are indebted to the following people for helpful discussions: Philipp Blum, David Braddon-Mitchell, Claudio Calosi, Marta Campdelacreu, Aurélien Darbellay, Paul Égré, Graeme A. Forbes, Akiko Frischhut, Francesco Gallina, Manuel García-Carpintero, Carl Hoefer, Miguel Hoeltje, John Horden, Dan López de Sa, Ned Markosian, José Martínez, Manuel Martínez, Giovanni Merlo, Kevin Mulligan, Bryan Pickel, Oliver Pooley, Simon Prosser, Pablo Rychter, Alessio Santelli, Gonçalo Santos, Thomas Sattig, Moritz Schulz, Amy Seymour, Albert Solé, Meghan Sullivan, Stephan Torre, Giuliano Torrengo, J. Robert G. Williams, Elia Zardini, Dean Zimmerman, and audiences at the LOGOS seminar, the LOGOS reading group on time, the *eidos* seminar, the LanCog seminar, the Cycle de conférences PhilEAs, the PERSP Metaphysics Seminar, the PERSP Workshop on the As and Bs in the Philosophy of Time, the second LOGOS Workshop on Vagueness and Metaphysics, the CUSO workshop on Experience and Reality, and audiences at the Universities of Milan, Neuchâtel and Tübingen.

We are also very much indebted to Otávio Bueno, editor of the *Synthese Library*, for his encouragement and support throughout the publication process.

Work on this monograph was partly funded by the Consolider-Ingenio project PERSP (CSD2009-00056) and the project *The Makings of Truth: Nature, Extent, and Applications of Truthmaking* (FFI2012-35026), both financed by the Spanish Ministry of Economy, by the European Commission's HORIZON 2020 Marie Skłodowska-Curie European Training Network *DIAPHORA*, under grant agreement H2020-MSCA-ITN-2015-675415, as well as by the Swiss National Science Foundation projects *Grounding - Metaphysics, Science, and Logic* (CRSII1-147685), *The Nature of Existence: Neglected Questions at the Foundations of Ontology* (100012-150289), and *The Metaphysics of Time and its Occupants* (BSCGI0_157792).

Contents

Chapter 1
Taking Tense Seriously

Abstract In this chapter we introduce the system of propositional tense logic that we will use throughout the book, clarify what it means to take tense seriously for the purposes of metaphysical enquiry, and clarify the contrast between dynamic and static conceptions of reality. In Sect. 1.1 we set out Arthur Prior's operator approach to tense and distinguish between the grammatical and the logical notions of tense, which latter calls for a systematic regimentation of ordinary language. In Sect. 1.2 we present axioms governing non-metric temporal operators, including non-standard operators of the form 'At *t*'. In Sect. 1.3 we use the notion of truth *simpliciter* to discern two broad types of metaphysical views that disagree on whether a complete description of reality requires the use of tense.

In this chapter, we first introduce the tense-logical regimentation of tensed language, familiar from the works of Arthur Prior, according to which, for metaphysical purposes, the tenses, simple and complex, can be represented by temporal operators, or combinations thereof, operating on present-tensed clauses. We call this approach to tense *the operator approach to tense*. We then develop a logical system whose axioms encode how these temporal operators behave, once the language is suitably regimented. For the purposes of this chapter, we focus on propositional tense logic, but will say more about how temporal operators interact with quantifiers in subsequent chapters. Lastly, we explicate what it means to take tense seriously in the context of metaphysical enquiry, and do so by appeal to the notion of truth *simpliciter*.

In the last decades, linguists and philosophers of language alike have adduced evidence suggesting that natural languages are best not conceived as having the kind of deep structure that the operator approach discerns (see e.g. Partee 1973; King 2003). Instead, these linguists and philosophers think that tenses in natural languages are better seen to function in ways similar to the ways in which pronouns do. We neither take issue with these claims nor with the evidence adduced in their support. Our concern is with the metaphysics of time; and it is an altogether independent question whether, if those claims are correct, natural languages are better suited to articulate competing ontological views than the regimented language of tense logic. Here we think that the regimented language of tense logic does extremely

F. Correia, S. Rosenkranz, *Nothing To Come*, Synthese Library 395,
https://doi.org/10.1007/978-3-319-78704-6_1

well and insist that it would be mistaken to conclude, from the fact, if it is a fact, that natural languages come with certain ontological commitments, that the right metaphysics of time must likewise incur them.

1.1 The Operator Approach to Tense

Tense is primarily a grammatical category. The *logical* notion of tense differs from the grammatical one; and unsurprisingly perhaps, we can win through to a grasp of the logical notion only once we logically regiment ordinary language.

To begin with note that according to surface grammar at least, in sentences like 'Three days ago, the world economy was collapsing' the adverbial phrase 'Three days ago' would seem to modify the past-tensed main clause. By contrast, in tense logic, the adverbial phrase 'Three days ago' is rendered as the temporal operator 'It was three days ago the case that' which shifts the circumstances against which to evaluate the embedded clause to those prevailing three days ago. Thus understood, in the case at hand, 'Three days ago' operates on a clause *in the present tense*. For, it does not take much thought to realise that, however grammatically awkward it may initially sound, the sentence 'It was three days ago the case that the world economy *is* collapsing' succeeds in saying something rather different from what the sentence 'It was three days ago the case that the world economy *was* collapsing' succeeds in saying: the latter, but not the former, is consistent with the thought that the present bonanza already began four days in the past of now; and plausibly, when we utter, in ordinary English, 'Three days ago, the world economy was collapsing' we wish to preclude this possibility. Accordingly, from a tense-logical perspective, the past tense in the embedded clause of that sentence is redundant (Prior 1967: 14; 2003: 13–14).

To the extent that tense-logical atoms are anyway always in the present tense, there is no need to represent the present tense by means of any operator. If we so wish, we can nonetheless introduce an operator 'Presently it is the case that' which allows us to prefix any tensed clause without change in truth-value. Any such operator will be redundant in the sense that 'Presently it is the case that φ' and φ are always tense-logically equivalent, irrespective of whether the clause φ is itself in the past, present or future tense – which is to say in tense-logical terms, irrespective of whether φ itself consists of a sentence prefixed by a temporal operator. Thus, on this reading, 'Presently it is the case that it was the case three days ago that the world economy is collapsing' will say no more and no less than 'It was three days ago the case that the world economy is collapsing'. Similarly, 'It was three days ago the case that, presently, the world economy is collapsing' will say no more and no less than 'It was three days ago the case that the world economy is collapsing' (Prior 1967: 14–15). The operator 'Presently it is the case that' is an idle wheel, which is why we will refrain from using it.[1]

According to the operator approach to tense, date terms can likewise function as temporal sentential operators; and we may get an even clearer picture of the difference between the grammatical and the logical notions of tense, once we reflect on statements like 'On May 10th, φ', made in everyday contexts (and not, say, in the context of a fiction). If we claim to know that May 10th is later than today, grammar dictates that we use the future tense in describing the goings-on on May 10th, and so should utter a sentence like 'On May 10th, the world economy *will be* collapsing'. Correspondingly, if we utter this sentence, competent speakers will typically understand our utterance to convey the information that May 10th is later than today. If by contrast we claim to know that May 10th is earlier than today, grammar dictates that we use the past tense instead, and so should utter a sentence like 'On May 10th, the world economy *was* collapsing'; and correspondingly, if we utter this sentence, our utterance will typically be understood to convey the information that May 10th is earlier than today. If we claim to know that May 10th is today, grammar dictates that we use the present tense in describing the goings-on on May 10th and so should rather utter a sentence like 'On May 10th, the world economy *is* collapsing'. It might therefore be expected that, similarly, whenever we utter 'On May 10th, the world economy *is* collapsing', our utterance is understood to convey the information that May 10th is today. But this is not what we find, and for good reason.

For suppose that this was indeed the case. What would we then be allowed to say whenever we claimed to know what the world economy on May 10th is like, but did *not* take ourselves to know whether or not that day is earlier than today, and so neither took ourselves to know that it is earlier than today, nor took ourselves to know that it is later than today, nor took ourselves to know that the 10th of May is today? There must surely be *some* way to communicate what we claim to know about the economic situation on May 10th which is unhampered by our ignorance about the temporal distance May 10th bears to today, zero or otherwise. Here, we cannot say 'On May 10th, either the world economy is, was or will be collapsing', in order to flag our ignorance. For, this disjunction is naturally understood as being consistent with May 10th being earlier than the collapse of the world economy, which we may after all have every reason to rule out. Nor, for that matter, can we say 'On May 10th, the world economy is, was and will be collapsing', because this would naturally be understood as being inconsistent with the collapse of the world economy on May 10th being the last we have to face, which latter we may have no reason at all to rule out. So if we want to ensure that we can communicate what May 10th is like eco-

[1] Just as in the case of the modal operator 'Actually', there is also a non-redundant operator of the same form which, as it were, seals off the clause it embeds from any effects that further tense-logical embedding might have, in the sense that 'Presently it is the case that φ' will prove tense-logically equivalent to 'Always, presently it is the case that φ' (see Kamp 1971). At a later stage, we will indeed invoke such an operator (see Chap. 7, Sect. 7.2).

nomically, even in situations in which we have no inkling as to the temporal relation that it bears to today, what can we say?

Once we take 'On May 10th' to function as a temporal operator that shifts the circumstances against which to evaluate the embedded clause to those prevailing on May 10th, we should use the present tense in formulating the embedded clause. Once 'On May 10th' is interpreted in this way, it is unsurprising that use of the present tense no longer conveys that May 10th is today. But on that same interpretation, if it is to be consistent, 'On May 10th, the world economy *will be* collapsing' will no longer convey that May 10th is later than today, and 'On May 10th, the world economy *was* collapsing' will no longer convey that May 10th is earlier than today – irrespective of what grammar might suggest or dictate. Instead, these sentences are then best taken to respectively convey that an economic collapse is in the future of May 10th, and that an economic collapse is in the past of May 10th. This illustrates once more the difference between the grammatical and logical notions of tense.[2]

In what follows we assume that the language to be used for the articulation, and discussion, of different metaphysical views about time and existence, is the regimented language of the operator approach to tense. Accordingly, we take tenses to be representable by means of temporal operators that shift the circumstances of evaluation of the clauses they embed, and presume that tense-logical atoms are always present-tensed. Once tensed language is regimented in this way, we can codify inferences that essentially rely on the operator-induced structure of sentences of that language. This codification is the task of *tense logic*.

1.2 Propositional Tense Logic

Temporal operators come in two varieties, metric and non-metric. Examples of non-metric temporal operators are 'Sometimes in the past', 'Sometimes in the future' and 'Always'. Examples of metric temporal operators are 'Four days ago' and 'Three days hence'. The tense logic we will introduce in what follows, and subsequently put to use, only employs non-metric temporal operators. Although all the proofs we devise will be couched in these terms, in informal philosophical discussion we will occasionally appeal to metric temporal operators. Apart from the more familiar non-metric operators, our logic will also deploy non-metric temporal operators of a less familiar type, *viz.* operators of the form 'At *m*', where '*m*' is a stand-in for terms that denote, or range over, times.

[2] This construal of 'On May 10th' is not the only possible one. We may instead regard 'On May 10th' as a syncategorematic part of the predication which latter is accordingly taken to attribute a *relation* between the world economy and May 10th. On this interpretation, we had better think of the predication itself as, logically speaking, untensed – in the sense in which, plausibly, 'is odd' in the mathematical statement 'The number 9 is odd' is, logically speaking, untensed. Once understood in this way, both 'On May 10th, the world economy *will be* collapsing' and 'On May 10th, the world economy *was* collapsing' will at best be misleading, but will anyway no longer be suitable means to convey anything about the relation that May 10th bears to today.

Temporal operators interact with the quantifiers in interesting ways; and as we shall see, in addition to minimal, theory-neutral principles prescribing this interaction, there are also further principles whose acceptance depends on one's ontological view. We will come back to these issues in the next chapter, where we will identify some such principles that we take to be minimal, and in Chaps. 4 and 5, where we will show which further such principles respectively hold on GBT, presentism and other contenders, and which ones do not. For now, we will just look at the propositional fragment of tense logic. Among the principles of propositional tense logic, there are some that encode assumptions about the structure, or topology, of time, e.g. that it is linear towards the past and towards the future.

We begin by explaining the use of the more familiar non-metric operators and the definitions, axioms and rules they will be taken to underwrite. Following established usage, we abbreviate 'Sometimes in the past' by 'P', 'Sometimes in the future' by 'F', 'Always in the past' by 'H' and 'Always in the future' by 'G'. Then P and F are definable as follows:

(D1) $P\varphi \equiv_{df} \neg H \neg \varphi$
(D2) $F\varphi \equiv_{df} \neg G \neg \varphi$

Thanks to (D1) and postulates to be laid down in what follows, H and P are duals: H is equivalent to ¬P¬ and P to ¬H¬. Likewise, G and F are duals in the same sense, thanks to (D2) and postulates we introduce below. We will assume that 'H' and 'G', as well as 'P' and 'F' as defined, underwrite the axioms and rules of minimal tense logic. Thus we have:

(A1) $\varphi \rightarrow HF\varphi$
(A2) $\varphi \rightarrow GP\varphi$
(A3) $H(\varphi \rightarrow \psi) \rightarrow (H\varphi \rightarrow H\psi)$
(A4) $G(\varphi \rightarrow \psi) \rightarrow (G\varphi \rightarrow G\psi)$

In other words, whatever is the case has always been going to be the case and is always going to have been the case; if it has always been the case that ψ is implied by some φ that has itself always been the case, ψ too has always been the case; and if it will always be the case that ψ is implied by some φ that itself will always be the case, ψ too will always be the case. We also have the following rules:

(R1) $\varphi / H\varphi$
(R2) $\varphi / G\varphi$

In other words, if φ is a theorem, then both that φ always has been the case, and that φ always will be the case, are likewise theorems. In addition to this minimal base, we lay down the following axioms concerning the topology of time:

(A5) $FP\varphi \rightarrow (P\varphi \vee \varphi \vee F\varphi)$
(A6) $PF\varphi \rightarrow (P\varphi \vee \varphi \vee F\varphi)$
(A7) $PP\varphi \rightarrow P\varphi$

(A5) and (A6) prescribe that, in the Kripke-models to be introduced in Appendix 1, time is linear both in the past direction and in the future direction, while (A7) prescribes that the accessibility relation between the times of these models is transitive. As such, none of these principles is entirely theory-neutral, even if (A5) and (A7) are far less contentious than (A6), given the much discussed view according to which indeterminism requires forward branching. However, it is unclear to what extent assuming all three of these principles prejudges any issues in GBT's favour: other views, such as presentism or eternalism, may likewise endorse them. Our principal concern is with the articulation of GBT. We will motivate later, in Chap. 7, why indeterminist GBT has no need for a branching conception of time.

We next introduce the temporal operators 'Always' and 'Sometimes'. They are respectively defined as follows:

(D3) Always, $\varphi \equiv_{df} (H\varphi \,\&\, \varphi \,\&\, G\varphi)$
(D4) Sometimes, $\varphi \equiv_{df} (P\varphi \vee \varphi \vee F\varphi)$

Accordingly, given (D1) and (D2), 'Always' and 'Sometimes' are duals: 'Always' is equivalent to '¬Sometimes¬' and 'Sometimes' to '¬Always¬'. Given (D3), the rules (R1) and (R2) yield the derived rule:

(RD1) φ / Always, φ

which we will frequently use. We can also establish

- Always, $(\varphi \to \psi) \to$ (Always, $\varphi \to$ Always, ψ)

Other standard tense-logical theorems and derived rules, for which we will not introduce labels, will be used in the proofs to follow. Thus, it can likewise easily be established that all of the following are theorems:

- (Always, $\varphi) \to \varphi$
- (Always, $\varphi) \to$ Always, Always, φ
- $\varphi \to$ Always, Sometimes, φ

Accordingly, 'Always' and 'Sometimes' behave like the modalities '□' and '◊' in C.I. Lewis' system S5. Similarly, we can establish the following theorems:

- $FF\varphi \to F\varphi$
- $H(\varphi \to \psi) \to (P\varphi \to P\psi)$
- $G(\varphi \to \psi) \to (F\varphi \to F\psi)$
- Always, $(\varphi \to \psi) \to$ (Sometimes, $\varphi \to$ Sometimes, ψ)
- $\varphi \to$ Sometimes, φ
- (Sometimes, Sometimes, $\varphi) \to$ Sometimes, φ
- (Sometimes, Always, $\varphi) \to \varphi$

as well as the following derived rules:

- $\varphi \to \psi$ / $H\varphi \to H\psi$
- $\varphi \to \psi$ / $G\varphi \to G\psi$
- $\varphi \to \psi$ / (Always, φ) $\to$ Always, ψ
- $\varphi \to \psi$ / $P\varphi \to P\psi$
- $\varphi \to \psi$ / $F\varphi \to F\psi$
- $\varphi \to \psi$ / (Sometimes, φ) $\to$ Sometimes, ψ
- $F\varphi \to \psi$ / $\varphi \to H\psi$
- $P\varphi \to \psi$ / $\varphi \to G\psi$
- (Sometimes, φ) $\to \psi$ / $\varphi \to$ Always, ψ

The proofs are routine in basic tense logic, which is why we omit them. Many of these theorems and derived rules will be useful in what follows. We will often indicate that we use such principles without making explicit exactly which ones.

As already indicated, we will also use non-metric temporal operators of the form 'At *m*', where '*m*' is used for terms that denote, or range over, times. Instead of introducing special constants and variables for times and imposing the grammatical requirement that the blank in 'At –' be filled only by such terms, we adopt as axiomatic the principle that whenever something holds at an entity, that entity must be a time. Let us use T as a predicate for times. The axiom can then be formulated as follows:

(A8) (At x, φ) $\to$ Tx

where T is itself supposed to obey the following axiom:

(A9) Tx $\to$ Always, Tx

That is to say, being a time is an eternal feature. Temporal operators of the sort under consideration are furthermore understood to underwrite the following axioms:

(A10) At x, $(\varphi \to \psi) \to$ (At x, $\varphi \to$ At x, ψ)
(A11) At x, $\neg\varphi \leftrightarrow$ (Tx & $\neg$At x, φ)

'At x, φ' in principle allows for the standard analysis in terms of 'Tx & Always, (x is present $\to \varphi$)'; but we do not wish to commit to this analysis. Notice that (A11) presupposes that 'At x, φ' does not entail 'x exists' given that 'Tx' does not.[3]

Temporal operators of this kind interact with the other non-metric temporal operators in the following ways:

(A12) (Always, φ) $\to$ (Tx $\to$ At x, φ)
(A13) (At x, φ) $\to$ Always, At x, φ

[3] In principle, we could define a corresponding operator 'AT x' such that 'AT x, φ' does entail 'x exists', by laying down 'AT x, $\varphi \equiv_{df}$ (x exists & At x, φ)'. Alternatively, we could start off by assuming such an operator and define 'At x' in its terms by laying down 'At x, $\varphi \equiv_{df}$ Sometimes, AT x, φ'.

In other words, what always holds, holds at all times; and what holds at a time, always holds at that time. Given (A12), temporal operators of the form 'At x' underwrite the following derived inference rule:

(RD2) φ / T$x \rightarrow$ At x, φ

Thanks to (A10) and (A8), by (RD2), we can then derive:

(RD3) $\varphi \rightarrow \psi$ / (At x, φ) $\rightarrow$ At x, ψ

From (A12) and (A13), we can derive:

(T1) (At y, φ) $\rightarrow$ (T$x \rightarrow$ At x, At y, φ)

Thanks to (A11), (RD3) and basic tense-logical principles, (A12) yields:

(T2) (At x, φ) $\rightarrow$ Sometimes, φ

Similarly, (A13) yields:

(T3) (Sometimes, At x, φ) $\rightarrow$ (At x, φ)

(T2) and (T3) jointly imply that the following theorem holds:

(T4) (At x, At y, φ) $\rightarrow$ At y, φ

Importantly, and with hindsight, we do *not* here presuppose that '$\forall x$ (T$x \rightarrow$ At x, φ)' logically entails 'Always, φ' or, equivalently, that 'Sometimes, φ' logically entails '$\exists x$(At x, φ)'. For, part of what is at issue between proponents of GBT and their opponents is whether there is a last time. If there is a last time, and if at that time, and at all preceding times, there are no human beings on planet Mars, it may still be true that there *will be* human beings on Mars. There will be further axioms we assume to govern operators of the form 'At m' that we can only introduce after moving to quantified tense logic (see Chaps. 2 and 3).

As in the case of the standard tense-logical principles, we will often mention that we make use of principles for the At-operators without making explicit exactly which ones.

1.3 Truth *simpliciter*

The propositional tense logic characterised above is available to everyone who accepts both that time is linear in both directions, and that the relation of precedence between times is transitive. Even if, as noted in the introduction to this chapter, many linguists and philosophers of language deny that natural language exhibits the kind of deep structure that allows tenses to be adequately represented by temporal operators, this finding need not worry the metaphysician, since for all that, the language of tense logic might still be a suitable tool to describe what temporal reality is like.

Metaphysicians do, however, disagree about whether we need a tensed language at all for this purpose – in other words, whether, as metaphysicians, we should take tense seriously. Thus, one may agree that an utterance of the sentence 'It is sunny but in 24 h it will rain' is an adequate means to record what reality is like in certain respects, and yet deny that tensed sentences of this kind are necessary in order to record what reality is like in those respects. Instead, one may hold that, to this end, it will do to utter the sentence 'It is sunny at *u*, but for some time *t* 24 h later than *u*, it rains at *t*', where '*u*' names the time of utterance. Such a sentence would seem to have a stable truth-value, and any utterance of it would accordingly seem to be true or false independently from when that utterance occurs. At least, this would be so, if on each occasion of utterance, there was guaranteed to be both a time for '*u*' to name and a time one day later than that time – and this may of course itself be a matter of dispute.

But even someone who is happy to concede that these auxiliary conditions are met may nonetheless deny that an utterance of the tenseless sentence 'It is sunny at *u*, but for some time *t* 24 h later than *u*, it rains at *t*' succeeds in capturing all that an utterance of the tensed sentence 'It is sunny but in 24 h it will rain', made at *u*, does capture. This need not imply that such a philosopher finds any fault with the following equivalence:

> 'It is sunny but in 24 h it will rain' is true at *u* $\leftrightarrow$ 'It is sunny at *u* & for some time *t* 24 h later than *u*, it rains at *t*' is true.

What she will anyway deny, however, is that the fact, if any, recorded by saying, at *u* or any other time, both that it is sunny at *u* and that for some time *t* 24 h later than *u*, it rains at *t*, is the very same fact recorded by saying, at *u*, that it is sunny but in 24 h it will rain.

Her opponent may (but need not) concede that the *proposition* expressed by an utterance, made at *u*, of 'It is sunny but in 24 h it will rain' differs from the *proposition* expressed by an utterance, made at *u* or any other time, of 'It is sunny at *u*, but for some time *t* 24 h later than *u*, it rains at *t*': one but not the other proposition has an unstable truth-value. But this does not resolve the disagreement. For her opponent will nonetheless insist that the *fact*, if any, recorded by an expression, at *u*, of the first proposition is the very same fact recorded by an expression, at *u* or any other time, of the second proposition.

To see more clearly what is at stake here, let us begin by distinguishing two thoughts:

(i) Among the tensed propositions, if any, there are many that are sometimes but not always true
(ii) In application to such tensed propositions truth is only ever relative to times and never absolute

Everyone should agree with the former. It does not follow that everyone should therefore agree with the latter. Of course, *if* in order for a proposition to be true relative to all times it is required that it always be true *(first assumption)*, and *if* being true absolutely implies being true relative to all times *(second assumption)*, then (ii) straightforwardly follows from (i). But neither of these additional assumptions is uncontroversial. Thus, on ontological views that countenance no future times, or no past times, or neither past nor future times, the first assumption will fail: φ may hold at all the times whose existence these views acknowledge, while 'Always, φ' nonetheless fails. (It merits emphasis that, as noted before, nowhere in the preceding sections did we presuppose that tense-logical operators allow for recapture in terms of quantification over times.) But even on views consistent with the first assumption, the second assumption might fail already because these views are inconsistent with the idea that what is true absolutely is always true; and it is this latter idea that will also be inconsistent with most views that reject the first assumption.

The contention that being true absolutely implies being true relative to all times does not imply that absolute truth is nothing but truth relative to all times. And it had better not imply that, for, if anything, absolute truth is not relative. Absolute truth is truth *simpliciter*, truth unqualified. So, if being true absolutely indeed implies being true relative to all times, then this will at best be so because relativisation to times has no further effect on absolute truths. In conjunction with the first assumption, (i) implies that, insofar as there are any tensed propositions at all, some tensed propositions are not true at all times because they are not always true. The second assumption accordingly implies that no such tensed proposition is ever true *simpliciter*. Accordingly, on a view such as this, the tensed proposition expressed by an utterance of 'It is sunny', if any, can never be true *simpliciter*, given only that it is not always sunny, and similarly *mutatis mutandis* for any tensed proposition that lacks a stable truth-value. This is something to which those metaphysicians who take tense seriously are bound to object.

It is natural to think that whatever facts there are – whichever way reality is 'in and of itself' – it is truths *simpliciter* that record such facts (Dummett 2006: 12). Thus, if tensed propositions that vary in truth-value across time are ever true *simpliciter*, then what facts there are changes with time or, at the very least, the way those facts are changes with time. For if the facts did not change in any way as time goes by, the very same propositions that once were true *simpliciter* should continue to be suitable means to record them. On the other hand, if the facts do change with time, only tensed truths *simpliciter* will be able to record them and their changes. Accordingly, a debate about which types of propositions qualify as truths *simpliciter* is a debate about whether or not reality itself ever changes.

We can now more clearly state what is at issue between those metaphysicians who take tense seriously and those who do not. Those metaphysicians who do not take tense seriously endorse *the static view*, i.e.

(STA) Always, $\forall p$ Always ($\mathcal{T}p \rightarrow$ Always, $\mathcal{T}p$)

where the operator '$\mathcal{T}$' is short for 'It is true *simpliciter* that' and underwrites the following principle:

$$\mathcal{T}\varphi \rightarrow \varphi$$

In other words, then, according to (STA), whenever something is a truth *simpliciter* it is always true *simpliciter*, and hence never changes its truth-value. By contrast, those metaphysicians who do take tense seriously will deny this and so instead endorse *dynamism*, i.e.

(DYN) Sometimes, $\exists p$ Sometimes ($\mathcal{T}p$ & Sometimes, $\neg \mathcal{T}p$)

In other words, sometimes something sometimes is a truth *simpliciter* without always being a truth *simpliciter*. To borrow a phrase of Wittgenstein's, reality is the totality of facts. To the extent that truths *simpliciter* record the facts that compose reality, (STA) accordingly corresponds to the idea that reality is static, while (DYN) corresponds to the idea that reality is dynamic: what reality was like in the past of now need not match what reality is like now, and what reality is like now need not match what reality will be like in the future of now.

If time neither extends into the future nor into the past, (STA) will be trivially true – and so (DYN) will be trivially false. On the other hand, the assumption that time extends into the future or past not only is highly plausible, it also is neutral between (STA) and (DYN). Accordingly, we shall throughout assume the following tense-logical axiom which can be seen to encode this assumption:

(A14) $P\top \vee F\top$, for $\top$ any chosen tautology

Note that our adoption of the operator approach to tense, and our corresponding assumption that the language to be used contains tensed clauses on which temporal operators operate, in no way prejudge the issue of whether tensed propositions that are not always true can ever be true *simpliciter*. Thus, opting for the static view according to which there are *no* tensed propositions apt to be true *simpliciter*, does not entail rejection of the suggestion, sketched above, that, say, in 'On May 10th, the world economy is collapsing' the phrase 'On May 10th' is a temporal operator that operates on a present-tensed clause and shifts the circumstances against which to evaluate the embedded clause to those prevailing on May 10th. For, even if such clauses, and the more complex sentences embedding them, may express tensed propositions, their truth may for all that still be relative to times and so not be absolute.

---

[4]This is one important respect in which $\mathcal{T}$ differs from Fine's 'In reality'-operator: unlike the latter, $\mathcal{T}$ is factive (Fine 2005: 268, 282).

Chapter 2
Existence, Quantification and Identity

Abstract In this chapter we introduce the distinction between permanentist and temporaryist ontologies and present a non-classical theory of unrestricted quantification and identity that is compatible with either type of view. We discuss and defuse a recent objection that temporaryism cannot accommodate unrestricted quantification. In Sect. 2.1 we use temporal operators and quantification in order to articulate the core tenets of permanentism and temporaryism, and show that static conceptions of reality are committed to permanentism. In Sect. 2.2 we observe that classical quantification theory favours permanentism, and for reasons of neutrality, replace it by a quantification theory that, jointly with corresponding axioms for identity, yields a positive free logic. In Sect. 2.3 we reject T. Williamson's argument meant to show that temporaryists should endorse the so-called temporal *being constraint*, lest they be accused of using restricted quantification when articulating their view.

In this chapter, we will begin by distinguishing two broad classes of ontological views that disagree about what there is: *permanentist* views and *temporaryist* views (cf. Williamson 2013). Among the members of each class, we will make further divisions that correspond to familiar theories of time and existence. As it will turn out, all friends of a static universe are permanentists but not the other way round; and all temporaryists are friends of a dynamic universe but not the other way round. Classical quantification theory is biased towards permanentist views. Consequently, we will adopt a weaker theory of quantification, so as to ensure a neutral starting ground. Throughout we will take quantification to be *unrestricted*: whenever we quantify, we quantify over absolutely everything. Recently it has been argued that unrestricted quantification favours permanentism. We respond to this charge and argue that temporaryism is perfectly compatible with the idea that quantification is unrestricted.

Throughout, we will use 'm', 'n' and variants thereof for terms, i.e. expressions that are either constants or variables. Existence will accordingly be taken to be defined in terms of quantification and identity as follows:

(D5) $\quad \mathrm{E}!m \equiv_{\mathrm{df}} \exists x(m = x)$

F. Correia, S. Rosenkranz, *Nothing To Come*, Synthese Library 395,
https://doi.org/10.1007/978-3-319-78704-6_2

where 'E!' abbreviates 'exists'.[1] In this spirit, we will assume as an axiom

(A15) $\forall x \mathrm{E!}x$

In addition, we assume

(A16) Sometimes, $\mathrm{E!}m$

Different metaphysical theories hold different views about what exists. But these differences are best not conceived as differences in the *notions* of existence and quantification they respectively employ (cf. Correia and Rosenkranz 2015b). Thus we take the quantifiers to be univocal, and hence likewise the predicate 'E!'. We will also presume that always, truths about what exists are truths *simpliciter*, so that we have

$$\mathrm{E!}m \rightarrow \mathscr{T}\mathrm{E!}m$$

(A15) and (A16) make use of the quantifiers and the concept of identity. We will give the details of the theory of quantification and logic of identity we here presuppose in Sect. 2.2.

2.1 Temporaryism *vs* Permanentism

Ontological views differ according to whether or not they take what exists to *permanently* exist. To properly express these different views, we must go beyond propositional tense-logic and see how temporal operators and quantifiers interact. However, as will become clear in due course, accounts of this interaction are not entirely theory-neutral, i.e. they depend, to some extent, on one's ontological views – just as certain features of the propositional tense-logic defined could already be seen not to be entirely neutral on the structure of time. As we shall highlight in the next section, even classical quantification theory itself is biased towards certain ontological views as opposed to others. To win through to a more neutral account of quantification that prejudges no issues, one must first know what those views are. It is therefore advisable to start off by introducing a broad classification of such ontological views.

Following Williamson (2013), let us distinguish between so-called *permanentist* views and so-called *temporaryist* views. The defining principle of permanentism is

(PER) Always, $\forall x$ Always, $\mathrm{E!}x$

[1] The variable x should of course be required to be distinct from m; and in a formally rigorous presentation we would specify which variable it is, e.g. the first variable distinct from m given a previously defined numbering of the variables. We will henceforth for the most part omit explicit mention of such provisos when we give definitions in such quantificational terms.

The defining principle of temporaryism, by contrast, is equivalent to the negation of (PER), i.e.

(TEM) Sometimes, $\exists x$ Sometimes, $\neg E!x$

Thus, according to (PER), it is never the case that what exists ever fails to exist, whereas according to (TEM) sometimes there exists something that sometimes does not exist. (PER) is reminiscent of the static view introduced in the previous chapter, i.e.

(STA) Always, $\forall p$ Always, $(\mathscr{T}p \rightarrow$ Always, $\mathscr{T}p)$

Indeed, (PER) follows from (STA). Thus, assume (STA). (STA) entails 'Always, $\forall x$ Always, $(\mathscr{T}E!x \rightarrow$ Always, $\mathscr{T}E!x)$'. By our two general assumptions about truth *simpliciter*, i.e.

$$\mathscr{T}\varphi \rightarrow \varphi$$
$$E!m \rightarrow \mathscr{T}E!m$$

we can derive 'Always, $\forall x$ Always, $(E!x \rightarrow$ Always, $E!x)$'. Thanks to (A16), we can then derive 'Always, $\forall x$ Sometimes, Always, $E!x$', and hence 'Always, $\forall x$ Always, $E!x$'. In other words, insofar as always, truths about what exists are truths *simpliciter*, everyone who believes in a static universe is *eo ipso* a permanentist: if facts never change, then neither do facts about what exists. Such a view, that derives its permanentism from its commitment to a static universe, is what is typically called *eternalism*.

However, the converse inference does not hold: one may believe that it is never the case that facts about what exists ever change, and nonetheless allow other kinds of facts to change with time. Indeed, we take the permanentist view expounded by Williamson (2013) to be of this latter kind. For, according to Williamson (2013), although it is never the case that some existent did not or will not exist, true propositions about what is concrete are nonetheless not always true, where, plausibly, such propositions are, whenever true, true *simpliciter* (see Chap. 5 for discussion).

Other views that accept (PER) but reject (STA) are conceivable. Thus, the classical *Moving Spotlight Theory* accepts a permanentist ontology, but insists that facts about what is present change with time. Yet another view that belongs to this class of views is the theory we expounded in Correia and Rosenkranz (2011, 2012). According to this view, although always everything, including the facts, always exists, facts nonetheless continually change in what might be called their 'tense-aspect': the fact that p will n days hence be the fact that n days ago, p, where at different times, different truths *simpliciter* capture the way this fact has evolved to be. Thus, on this view, reality changes with time even if it does not change in what it contains, including the facts; the facts it contains simply age. We may call all these views that accept (PER) but reject (STA) in favour of (DYN), varieties of *dynamic permanentism*. Eternalism may correspondingly be dubbed *static permanentism*.

Accordingly, we must systematically distinguish between significantly tensed *quantification* and significantly tensed *predication*. To say that quantification is significantly tensed is to say that a proposition of the form '$\exists x\varphi x$' can be true *simpliciter* without always being true, even in cases in which φ denotes a property that a thing can only ever have if it has that property throughout its existence, e.g. the proposition that $\exists x(x = \text{Fabrice})$. By contrast, to say that φ is significantly tensed is to say that a proposition of the form 'φa' can be true *simpliciter* and yet sometimes when *a* exists be false, e.g. the proposition that Sven is sleepy. While dynamic permanentists and static permanentists agree that quantification is not significantly tensed, dynamic permanentists affirm, while static permanentists deny, that some predications are significantly tensed.

Temporaryists are bound to reject (STA) and to opt for (DYN) instead. Thus, temporaryists believe in a dynamic universe. But as the availability of dynamic permanentism testifies, not all believers in (DYN) need therefore be temporaryists. We can further subdivide temporaryist views depending on which of the following two theses they accept:

(TEM_F) Sometimes, $\exists x\text{F}\neg\text{E!}x$
(TEM_P) Sometimes, $\exists x\text{P}\neg\text{E!}x$

Presentism, as this view is traditionally understood, accepts both these theses. For proponents of this view, always, the present time, conceived as concrete rather than abstract, never existed in the past and never will exist in the future. Presentism implies more than just acceptance of (TEM_F) and (TEM_P); and we will give a more comprehensive characterisation of the view in Chap. 5 (see also Correia and Rosenkranz 2015b). GBT differs from presentism and permanentism, in both its varieties, in that it accepts (TEM_P) and rejects (TEM_F). A more detailed characterisation of GBT will be given in Chap. 4. By contrast, a view according to which (TEM_F) holds but (TEM_P) fails to hold, has not found any supporters in the literature. We agree that such a view is highly implausible, which is why we will ignore it here and in the remainder.

As indicated in the introduction, we are on the defensive. Just as we are not trying to show that GBT is superior to other forms of temporaryism, we are not undertaking the attempt to argue that temporaryism is superior to permanentism. We merely intend to show that GBT makes good sense and that its core is not refuted, or challenged, by any extant *a priori* argument. Since GBT is a version of temporaryism, defending temporaryism is a good place to start. In the next section, we will present a proof of permanentism based on classical quantification theory and argue that since classical quantification theory thus proves to have a metaphysical bias, it should not be taken as basic but be replaced by a weaker theory. This does not in turn prejudge any issues, because conjoining the weaker theory with (PER) straightforwardly yields classical quantification theory. The proof is not new, but may be unfamiliar to some readers, which is why we present it in what follows. In the next but one section, we will then critically review an argument, recently advanced by Williamson (2013), to the effect that appeal to unrestricted quantification puts temporaryism on the spot.

2.2 Quantification Theory and the Logic of Identity

Recall the definition of 'exists' and the axioms (A15) and (A16) which we already postulated in the introduction to this chapter:

(D5) $E!m \equiv_{df} \exists x(m = x)$
(A15) $\forall x E!x$
(A16) Sometimes, $E!m$

The universal quantifier and the existential quantifiers are duals. We take the universal quantifier as basic and define the existential quantifier in its terms in the usual way. The universal quantifier underwrites the following axioms and the following rule:

(A17) $\forall x(\varphi \rightarrow \psi) \rightarrow (\forall x\varphi \rightarrow \forall x\psi)$
(A18) $\varphi \rightarrow \forall x\varphi$ (with x not free in φ)
(R3) $\varphi / \forall x\varphi$

From this we obtain classical quantification theory by adding the axiom

(A_{CQT}) $\forall x\varphi \rightarrow \varphi[m/x]$

where '$\varphi[m/x]$' is the result of freely replacing each free occurrence of x in φ by m. Note that given (A_{CQT}) and (A15), (A16) is derivable.

However, given (A15), (R3) and (RD1), (A_{CQT}) entails a substantial ontological thesis, *viz.* permanentism. This is shown by the following textbook proof:

(1)	$\forall x E!x \rightarrow E!x$	by (A_{CQT})
(2)	$E!x$	from 1 by (A15)
(3)	Always, $E!x$	from 2 by (RD1)
(4)	$\forall x$ Always, $E!x$	from 3 by (R3)
(5)	Always, $\forall x$ Always, $E!x$	from 4 by (RD1)

Temporaryists are therefore forced to reject (A_{CQT}) or some of the other postulates used in the proof. A natural move, which we will follow here, is to keep these other postulates, and to replace (A_{CQT}) by the weaker principle

(A19) $\forall x\varphi \;\&\; E!m \rightarrow \varphi[m/x]$

where '$\varphi[m/x]$' is the result of freely replacing each free occurrence of x in φ by m. Conjoined with (PER), (A19) and (A16) yield (A_{CQT}), as the following proof shows:

(1)	$\forall x\varphi$ & $E!m \rightarrow \varphi[m/x]$	(A19)
(2)	$\forall x$ Always, $E!x$	by (PER)
(3)	$E!m \rightarrow$ Always, $E!m$	from 2 by (A19)
(4)	Sometimes, $E!m \rightarrow$ Sometimes, Always, $E!m$	from 3 by tense logic
(5)	Sometimes, Always, $E!m$	from 4 by (A16)
(6)	$E!m$	from 5 by tense logic
(7)	$\forall x\varphi \rightarrow \varphi[m/x]$	from 1, 6

So, we do not prejudge any issues by laying down (A19) and (A16) as axioms instead of (A_{CQT}): all we are saying is that if (PER) holds, it does not hold as a matter of logic alone.

The addition of (A19), instead of (A_{CQT}), yields a so-called *free logic*. The choice between classical 'unfree' logic and free logic primarily turns on the interpretation of the singular terms of a given language and not primarily on the interpretation of the quantifiers themselves. There is thus no evident reason to presume that temporaryists and permanentists *eo ipso* imbue the quantifiers with different meanings, which would have the undesirable effect that they might be accused of talking past each other, rather than disagreeing about the domain and about the question whether it varies over time. Any accusation that the parties disagree about the meaning of the quantifiers rather than their extension thus stands in need of substantive argument.

This completes our theory of quantification. We will now introduce the identity predicate '=' as a further logical constant. We take '=' to obey the following standard axioms:

(A20) $x = x$
(A21) $x = y \rightarrow (\varphi \rightarrow \varphi[y//x])$

where '$\varphi[y//x]$' is the result of freely replacing zero or more free occurrences of x in φ by y. Thanks to (A16), we can show that if φ is a theorem, then so is '$\varphi[m/x]$', where '$\varphi[m/x]$' is the result of freely replacing each free occurrence of x in φ by m.

Proof: Suppose that φ is a theorem. Then by (RD1) and (R3), '$\forall x$ Always, φ' is also a theorem. By (A19), then, 'm exists $\rightarrow$ Always, $\varphi[m/x]$' is a theorem. Hence, by tense-logical reasoning, '(Sometimes, m exists) $\rightarrow$ Sometimes, Always, $\varphi[m/x]$' is a theorem. By (A16), then, 'Sometimes, Always, $\varphi[m/x]$' is a theorem. By tense-logical reasoning, we infer that '$\varphi[m/x]$' is a theorem. □

From this fact, one can infer that the following generalisations of (A20) and (A21) are theorems:

- $m = m$
- $m = n \rightarrow (\varphi \rightarrow \varphi[n//m])$

where '$\varphi[n//m]$' is the result of freely replacing zero or more free occurrences of m in φ by n. Other axioms with free variables give rise to similar generalisations.

Notice that thanks to the generalisations of (A20) and (A21) and (RD1), identity is also *eternal* in the sense that

– $m = n \rightarrow$ Always, $m = n$

is a theorem. Thanks to this and tense-logical reasoning, one can establish that distinctness is eternal in the same sense, i.e. that the following is also a theorem:

– $m \neq n \rightarrow$ Always, $m \neq n$

By contrast, neither identity nor distinctness will be assumed to be *existence-entailing*. The reason is immediate in the case of identity: if we accepted '$m = m \rightarrow$ m exists' as a theorem, since '$m = m$' is a theorem, we would have to accept 'm exists' as a theorem, and so we would after all be committed to (PER). That '$m = m$ $\rightarrow$ m exists' fails to hold shows that our logic is a *positive* free logic. In the case of distinctness, we can argue as follows. Assume for *reductio* that distinctness is existence-entailing. Then since '$m \neq n \rightarrow$ Always, $m \neq n$' is a theorem, '$m \neq n \rightarrow$ Always, m exist' is also a theorem. But Sven $\neq$ Fabrice, and yet, given GBT, Sven came into existence at some point in the past, and so does not always exist.

With this quantification theory and this logic of identity being in place, we can now already say something about how the quantifiers and '=' interact with the temporal operators introduced in the preceding chapter. To begin with, we will assume that the quantifiers interact with operators of the form 'At x' in the following way:

(A22) $\forall x(\mathrm{T}x \rightarrow \mathrm{At}\ x, \varphi) \rightarrow \varphi$ (with x not free in φ)

The intuitive rationale for (A22) is that, among the times within the range of the universal quantifier, there is the present time, and whatever holds at the present time, presently holds. Given (A11), (A22) yields

(T5) $\varphi \rightarrow \exists x\, \mathrm{At}\ x, \varphi$ (with x not free in φ)

(A22) also yields

(T6) $\exists x \mathrm{T}x$

To see this, simply let φ be any contradiction in which x is not free. In addition to (A22), we assume

(A23) $\mathrm{T}x \rightarrow \mathrm{At}\ x, \mathrm{E!}x$

i.e. the axiom that each time exists at itself, which seems eminently plausible, for whenever else should a time exists if it does not exist at itself?

Note that thanks to (A12) and the fact that identity and distinctness are both eternal, the following are likewise theorems

- $m = n \rightarrow (\mathrm{T}m' \rightarrow \text{At } m', m = n)$
- $m \neq n \rightarrow (\mathrm{T}m' \rightarrow \text{At } m', m \neq n)$

Lastly, we assume the following axiom

(A24) $(\text{At } x, \forall y\varphi) \leftrightarrow \text{Always}, \forall y \text{ At } x, (\mathrm{E}!y \rightarrow \varphi)$

where x and y are distinct variables. The intuitive rationale of (A24) is this: the expression 'Always, $\forall y$' acts somewhat like a quantifier whose range would be the set of all past, present and future objects; (A24) then encodes the fact that the range of '$\forall y$' at a given time is the restriction of the 'range' of the pseudo-quantifier 'Always, $\forall y$' to what exists at that time.

In which other ways temporal operators and quantifiers interact depends to a large extent on one's metaphysical theory. In Chap. 4 we will elaborate which further principles governing their interaction are licensed by GBT. GBT is a variety of temporaryism. In the remainder of this chapter, we discuss whether there is any reason to doubt that once temporaryism is understood in accordance with our definition of 'E!', i.e. (D5), it coheres with the thought that the quantifiers here introduced are unrestricted.

2.3 Unrestricted Quantification and the Temporal *Being Constraint*

We said in the introduction to this chapter that we take quantification to be unrestricted. We take this assumption to be consistent with temporaryism, i.e. with

(TEM) Sometimes, $\exists x$ Sometimes, $\neg \mathrm{E}!x$

Given the definition of 'E!', i.e.

(D5) $\mathrm{E}!m \equiv_{\mathrm{df}} \exists x(m = x)$

(TEM) is equivalent to

(1) Sometimes, $\exists x$ Sometimes, $\neg\exists y(x = y)$

Timothy Williamson has recently argued that the combination of (1) with the claim that the quantifiers involved are unrestricted is problematical, at least once it is seen in the context of further constraints and desiderata.

Temporaryism is for the temporal case what contingentism is for the modal case: contingentists claim that possibly something is possibly nothing (Williamson 2013: 2, 4). In Chap. 4 of his 2013 book, Williamson discusses whether, and if so in what form, contingentists should endorse the modal *being constraint*, i.e. the suitable generalisation to *n*-place predicates, for all *n*, of

(2) $\Box\forall x\, \Box(\Phi x \rightarrow \exists y(x = y))$

where permissible substitution instances of (2) are understood to be those that result from replacing 'Φ' in (2) by a (monadic) predicate (Williamson 2013: 148–58).[2] Since his discussion of contingentism is meant to likewise apply, *mutatis mutandis*, to temporaryism (Williamson 2013: 150), we will here critically review what Williamson has to say in that chapter, by implication as it were, about the relation between temporaryism and the *temporal* being constraint, i.e.

(3) Always, $\forall x$ Always, $(\Phi x \rightarrow \exists y(x = y))$

where permissible substitution instances of (3) are again understood to be those that result from replacing 'Φ' in (3) by a (monadic) predicate.[3]

If (3) is to be acceptable to temporaryists, the condition on permissible substitution instances of (3) just mentioned cannot be of a purely syntactic character. We can easily stipulate that for any variable (or singular term) *v*, '*v* is wanting' means the same as '$\neg\exists v'(v = v')$', for some variable *v'* distinct from *v*. It accordingly transpires that temporaryists *qua* temporaryists are bound to deny that (3) holds for every replacement of 'Φ' by what is, *according to purely syntactic criteria*, a predicate. For, if (3) holds with such generality, it holds for 'is wanting' and so entails

(4) Always, $\forall x$ Always, (x is wanting $\rightarrow \exists y(x = y)$).[4]

Given that '*x* is wanting' means the same as '$\neg\exists y(x = y)$', then thanks to the logical validity of '$(\neg\varphi \rightarrow \varphi) \rightarrow \varphi$', (4) entails the negation of (1). And *modulo* (D5), the negation of (1) is equivalent to the thesis of permanentism, i.e.

(PER) Always, $\forall x$ Always, E!x

---

[2]The intended generalisation to polyadic predicates follows the pattern '$\Box\forall x\Box\forall x'\ldots\ \Box(\Phi xx'\ldots \rightarrow \exists y(x = y)\ \&\ \exists y(x' = y)\ \&\ \ldots)$'.

[3]The intended generalisation to polyadic predicates follows the pattern 'Always, $\forall x$ Always, $\forall x'\ldots$ Always, $(\Phi xx'\ldots \rightarrow \exists y(x = y)\ \&\ \exists y(x' = y)\ \&\ \ldots)$'. Henceforth we will ignore this generalisation. Similarly, we will omit the qualification 'temporal' when referring to (3) as 'the being constraint'.

[4]Notice that irrespective of its logically more complex *definiens*, 'is wanting' is syntactically a predicate.

Accordingly, from the temporaryists' point of view, the acceptability of (3) is conditional on whether a stronger condition on its permissible substitution instances can be imposed. What should independently be clear is that, on their view, predicates that yield a true instance of (3) fail to be predicates for which the corresponding instance of (5) holds:

(5) Always, $\forall x$ Always, Φx

As we have seen in the previous section, the existence predicate is just such a predicate: for the temporaryist, it underwrites the relevant instance of (3) but not the relevant instance of (5). On any given choice of 'Φ', the conjunction of (3) and (5) all too obviously contradicts the temporaryists' key claim, i.e. (1) (cf. Williamson 2013: 154, 156; see also the next chapter for further discussion). However, the requirement that substitution instances for 'Φ' which validate (3) must not underwrite the corresponding instances of (5), evidently provides no sufficient condition, as aptly illustrated by 'is wanting': given (A15), 'Always, $\forall x$ Always, x is wanting' is clearly incoherent, while, given (1), (4) still fails.

The same example shows that temporaryists should agree with Williamson that the purely syntactic distinction between *positive* and *negative* predications is ill-suited to subserve any systematic restriction on the permissible substitution instances of (3) in line with their view (Williamson 2013: 157). Syntactically, 'is wanting' is positive rather than negative.

For all that has here been said, temporaryists may likewise be well-advised to agree with Williamson that, were they to accept (6), their view would 'slide into [permanentism] unless they complicate[d] [the] logic [of the λ operator] in awkward ways' (Williamson 2013: 188):

(6) Always, $\forall x$ Always, $(\lambda y(\Phi y)x \rightarrow \exists y(x = y))$.

Here, '$\lambda y(\Phi y)x$' is short for 'x is such that it Φs'. Any such move on the temporaryists' part would only make sense on assumption that 'Φx' does not in turn entail '$\lambda y(\Phi y)x$', for 'x is wanting' is of the form 'Φx' and (4) fails if (1) holds (cf. Williamson 2013: 158). According to Williamson, however, the latter assumption comes at high theoretical costs. Let us simply suppose without argument that, in this, he is right.

For all that, however, temporaryists may just take these observations to show that they have good reason to abandon the project of retaining (3) (and (6)) in any systematically restricted version that still is interestingly, and so non-trivially, general. The 'powerful prima facie attraction' of (3) that Williamson perceives (Williamson 2013: 148) might well be resisted in the light of examples such as 'is wanting'; and its felt force might be traced back to a limited range of favourable examples that tend to first come to mind (e.g. 'has mass', 'is coloured' or 'digests'), while unfavourable examples such as 'is wanting' do not.

Thus, the question arises why temporaryists should at all be obliged to come up with any interesting version of the being constraint that is neither geared to a particular list of examples nor entirely uninformative, amounting to the triviality that (3) holds for all those substitution instances of 'Φ' for which it does hold.[5] Williamson claims to have an argument for thinking that temporaryists are so obliged. In the following passages, which are here adapted to fit the temporal case, quite in line with what is intended, he argues

> Denials of [3] sound like failures to grasp the radical nature of unrestricted quantification. (Williamson 2013: 149)
>
> Indeed, the being constraint makes [temporaryism] more wholehearted. Without it, [temporaryism] looks ambivalent: the supposed counterexamples to the being constraint are pictured as casting enough of a [temporally] modal shadow on circumstances from which they are absent to bear properties and relations without being present themselves. Although such spatial pictures are easily imaginable in themselves, they betray the [temporaryist] when applied to the being constraint, since they represent the supposed counterexamples to it as merely elsewhere, within the range of an unrestricted quantifier and therefore something in the relevant sense, and merely out of range of a quantifier restricted to local things. They give comfort only to those who have failed to grasp how radical is the nothingness required of counterexamples to the being constraint. Hard-line [temporaryists] will accept the constraint. (Williamson 2013: 156)
>
> If [temporaryists] insist that [something may sometimes] fall under a predicate and yet be nothing, they face the charge that they are unserious about their own [temporaryism], because they are tacitly restricting the quantifier 'nothing'. (Williamson 2013: 188)

In the light of the foregoing observations, this is a curious line of argument. For, it is quite unclear what notions of predicate and property are involved. There are two alternatives. Either the argument just quoted operates with the purely syntactic notion of a predicate, and correspondingly conceives of the property denoted by such a predicate in terms of the condition something has to meet in order to fall under that predicate – in which case, as the definability of 'is wanting' shows, the argument is directed against temporaryism as such.[6] Its conclusion, that

[5] In the contingentists' case, the corresponding trivialisation could be formulated by saying that (2) holds only for those (predicates expressing) properties that are *existence-entailing in the purely modal sense*. Although temporaryists may likewise be happy to say that (3) holds for all (predicates expressing) such existence-entailing properties, the relevant trivialisation of (3) allows for a *prima facie* more general claim: trivially, (3) holds for all substitution instances of 'Φ' that something *actually only ever* satisfies when it exists, even if something *could* satisfy them without existing. Let '@' rigidly refer to the actual world. Then always for all x, always, if x is identical to @, x exists. However, arguably albeit controversially, @ is such that possibly it is self-identical but does not exist.

[6] Note that temporaryists do not claim that sometimes something satisfies the condition denoted by 'is wanting'. Rather, they merely claim that sometimes something *sometimes* satisfies that condition. This has a bearing on other things Williamson says in the chapter under discussion. Thus, in an interlude, Williamson observes that contingentists – and by extension, temporaryists – will deny that

(i) $\Phi a \rightarrow \exists y(a = y)$

is logically valid. His subsequent discussion would seem premised on the assumption that in order to argue their case, contingentists (temporaryists) appeal to empty names, of which fictional and temporaryists had better endorse the being constraint, would then at best be misleading, since it would reduce to the recommendation that they had better abandon their view, lest they be subject to the charge of failing 'to grasp the radical nature of unrestricted quantification'. As we shall argue in due course, though, any such charge would be misguided.

Alternatively, Williamson's argument may have to be understood as presupposing more restrictive, hitherto unexplained readings of 'predicate' and 'property' that respectively exclude 'is wanting' and *being wanting* from their range – in which case these readings should be made explicit. For, until they are out in the open, it is hard to gauge what dialectical force his argument really has. In particular, if they are such that the argument ultimately recommends acceptance of (6) – and hence are more restrictive readings only on condition that the implication from 'Φx' to '$\lambda y(\Phi y)x$' fails – any subsequent argument to the effect that this condition can only be met at high theoretical costs, immediately casts doubt on the cogency of the claim that temporaryists *qua* temporaryists had better accept (6). For then, for fear of such costs, commitment to (4) may be difficult to avert once (6) is accepted, since as we have seen, 'x is wanting' is syntactically of the form 'Φx'.

In any case, however, it is not the temporaryists' task to come up with a suitably systematic reading of 'predicate', in order to bolster an argument to the effect that they ought to accept all those substitution instances of (3) that result from replacing 'Φ' by predicates as thus understood. This would seem to be Williamson's own task – unless, that is, he can insist that his argument is dialectically effective even if it is understood to operate with the purely syntactic notion of a predicate. So let us first ask whether this is so, and hence whether, by endorsing that sometimes something sometimes is wanting, temporaryists fail 'to grasp the radical nature of unrestricted quantification'.

As we shall see and elaborate in Chap. 4, according to C. D. Broad, 'the sum total of existence is always increasing' (Broad 1923: 66–67). This slogan is only uncharitably construed as implying the absurdity that the sum total of what there is, some-

mythological names provide stock examples. However, they need have no problem with the idea that all instances of (i) are true. What they will object to is the claim that all instances of (i) necessarily (always) hold. However, if (i) was logically valid, all its instances *would* necessarily (always) hold. Proponents of GBT, for example, will accept (ii) but, unlike their permanentist opponents, deny (iii):

(ii) Socrates is wanting $\rightarrow \exists y(\text{Socrates} = y)$
(iii) Always in the past, (Socrates is wanting $\rightarrow \exists y(\text{Socrates} = y)$).

The truth of (ii) implies that 'Socrates' is not an empty name, while the falsity of (iii) in no way requires that whenever Socrates was wanting, 'Socrates' was nonetheless among the names then available, even if it then was empty. Thus, temporaryists may concede that 'we should distrust attempts to use fictional or mythological names to refute metaphysical or logical theses' (Williamson 2013: 153), but ask back why they should be described as having ever been tempted to undertake such attempts in order to argue their case. Similar considerations apply to contingentism, Williamson's explicit target.

times in the past, contained less. For, sums are never anything less than what they ever sum. Rather, what Broad intends to say here is that always there are things that, always in the past, were not among the things that then existed. Does saying so betray a failure to grasp the nature of unrestricted quantification, prompted by misleading pictures that draw on a permanentist ontology of things distributed across distinct temporal whereabouts?

As we shall see, on Broad's construal of the view, GBT entails that 'there is no such thing as ceasing to exist' (Broad 1923: 69), and so – contrary to what presentists will want to say – that always everything always in the future will be something. Accordingly, on this view, being something does not entail being present. For example, according to Broad, there are things that are not present but past and therefore are not, in any reasonable sense, *located* at the present time. In the light of this further claim, saying that always there are things that *always in the past were nothing*, would betray Broad's failure to grasp the nature of unrestricted quantification, if his sole reason for saying so was that always there are things that, *at any past time, are not located at that time*. For then 'being located at the present time' would tacitly, and quite inappropriately, restrict the tense-logically embedded quantifier 'nothing' – just as Williamson would seem to suggest.

However, this is quite clearly not Broad's reason. Thus, he is not at all concerned to profess truisms such as that always there is a present time. Rather, as we shall see more clearly in Chap. 4, his controversial idea is that always what there unrestrictedly is comprises more than what, at any earlier time, was what there then unrestrictedly was. This idea of continual growth is in no way undermined by the observation that whenever we quantify over what there unrestrictedly is, the range of our quantifier includes things that it would not include were we to restrict its range to things that, at a given past time, were present or past. Any such restriction merely traces what, at that past time, was – but now no longer is – the boundary of absolutely everything.

It is very plausible to assume that Williamson is under no illusion in this regard, and so it is equally plausible to assume that in the passages previously quoted, he presupposes suitably restricted readings of 'predicate' and 'property'. Yet, he himself does not explicate any such readings that would, at the same time, vindicate his conclusion that temporaryists had better endorse the being constraint as thus understood. In a way this is unsurprising, because he himself subsequently argues that temporaryists cannot without problems avail themselves of any such readings. What is surprising is that Williamson fails to see that the latter diagnosis, if correct, bereaves his argument to that very conclusion of any dialectical force.

Summarising his discussion, Williamson writes:

> [Temporaryists] are in a tricky position. If they insist that [something may sometimes] fall under a predicate and yet be nothing, they face the charge that they are unserious about their own [temporaryism], because they are tacitly restricting the quantifier 'nothing' [...]. If they agree that falling under a predicate entails being something, they slide into [permanentism] unless they distinguish not falling under a predicate from falling under a negative predicate, which is best done by means of something like the λ operator. If they introduce the λ operator, they still slide into [permanentism] unless they complicate its logic in awkward ways. (Williamson 2013: 188)

On condition that the second horn can be established, it follows that, in the light of desiderata such as simplicity, temporaryists have no convincing way to systematically improve upon the purely syntactic constraint on permissible substitution instances of (3) – at least no convincing such way that would make the resulting version of the being constraint at once interestingly general and such as to stably cohere with their view. Let that be so.

As we have just seen in relation to Broad's view, if understood to operate with the purely syntactic notion of a predicate, the quoted argument proffered for the first horn fails to deliver. For, a reading of the temporaryists' core thesis in terms of unrestricted quantification is perfectly coherent, where this thesis is readily expressible by use of 'is wanting'. Suppose instead that the argument for the first horn is premised on the availability of an alternative, not purely syntactic and yet still sufficiently general notion of a predicate, according to which 'is wanting' does not qualify as such. Then since the second horn implies that temporaryists cannot coherently avail themselves of such a notion without incurring considerable theoretical costs, the argument can easily be resisted on the basis of that very observation. In other words, if there indeed is theoretical pressure to accept the implication from 'Φx' to '$\lambda y(\Phi y)x$', and if, for temporaryists, the only way to insulate the being constraint from easy refutation by counterexamples such as 'is wanting' is by rejecting that implication and regarding the being constraint as properly stated by (6), then, for them, there is equal pressure to reject the being constraint. For, it is agreed that temporaryists have a principled reason to reject (4), even on a reading of the quantifiers as unrestricted. Hence one cannot both insist that there is such pressure and in the same breath accuse the temporaryists' rejection of the being constraint of being unprincipled or betraying the nature of unrestricted quantification.

At a more abstract level, the dialectical situation would seem to be this: Williamson seeks to persuade temporaryists that satisfaction of '$\lambda y(\Phi y)x$' demands existence, for any Φ, and then argues that there are sound simplicity-based reasons to regard satisfaction of 'Φx' to be at least as demanding as satisfaction of '$\lambda y(\Phi y)x$'. But temporaryists who, *qua* temporaryists, think that for certain Φ such as 'is wanting', satisfaction of 'Φx' does not demand existence, will take the latter argument to show that there are sound simplicity-based reasons to regard satisfaction of '$\lambda y(\Phi y)x$' to be as *un*demanding as satisfaction of 'Φx', thereby undermining the contention that '$\lambda y(\Phi y)x$' demands existence, for any Φ.

To sum up, granted that it can be established that there are sound reasons to accept the implication from 'Φx' to '$\lambda y(\Phi y)x$' – and we have done nothing here to either confirm or disconfirm the claim – the lesson is not that temporaryists 'are in a tricky position', but at best that they should reject the being constraint while foregoing any attempt to resuscitate, in non-trivial ways, a restricted, yet still sufficiently general version of it.

Chapter 3
Temporal Relations

Abstract In this chapter we introduce the relations of temporal location and precedence, critically review McTaggart's conception of the existential import of these relations, and devise axioms governing them that are acceptable to permanentists and temporaryists alike. In Sect 3.1 we critically review McTaggart's characterisation of the B-series according to which B-series relations are permanent, distinguish between two relevant senses of 'permanent' as applied to relations, and show that depending on whether such relations are taken as existence-entailing, on either reading, temporaryists can consistently avail themselves of such relations. In Sect. 3.2 we clarify the relation of temporal location by laying down axioms for it, taking into account things in time of various kinds. In Sect. 3.3 we first provide axioms for the relation of precedence amongst times, and then use it to define a relation of precedence for things in time quite generally.

In this chapter, we introduce two basic temporal relations, *location* and *precedence*, and lay down a number of axioms governing them. The former relation allows us to define a third temporal relation, *viz.* that of *contemporaneity*. These relations order things in a so-called B-series. We start by critically reviewing a passage from McTaggart (1927) where he characterises B-series relations as being *permanent* and from this infers that things belonging to the B-series always exist, and hence that, in application to them, the thesis of permanentism holds. This would make the relations of location and precedence ill-suited to serve as tools for the proper articulation of temporaryist ontologies. As we shall argue, however, there are two possible senses in which B-series relations might be said to be permanent: one according to which they are *eternal* and another according to which they are *rigid*. Depending on whether they take contemporaneity and precedence to be existence-entailing (in the temporal sense) – and so to obey Williamson's (temporal) *being constraint* – temporaryists have a reason to accept one rather than the other option. Neither combination of views leads to permanentism.

F. Correia, S. Rosenkranz, *Nothing To Come*, Synthese Library 395,
https://doi.org/10.1007/978-3-319-78704-6_3

3.1 McTaggart on the B-Series

When McTaggart introduced his famous distinction between the A- and the B-series of time, he had the following to say about the latter:

> Each position [in time] is [e]arlier than some and [l]ater than some of the other positions. To constitute such a series there is required a transitive asymmetrical relation, and a collection of terms such that, of any two of them, either the first is in this relation to the second, or the second is in this relation to the first. We may take here either the relation of 'earlier than' or the relation of 'later than', both of which, of course, are transitive and asymmetrical. If we take the first, then the terms have to be such that, of any two of them, either the first is earlier than the second, or the second is earlier than the first. [...] [These] distinctions [...] are permanent [...]. If *M* is ever earlier than *N*, it is always earlier. [...] The series of positions which runs from earlier to later, or conversely, I shall call the *B* series. The contents of any position in time form an event. The varied simultaneous contents of a single position are, of course, a plurality of events. [...] If *N* is ever earlier than *O* and later than *M*, it will always be, and has always been, earlier than *O* and later than *M*, since the relations of earlier and later are permanent. *N* will thus always be in a *B* series. [...] That is, it always has been an event, and always will be one, and cannot begin or cease to be an event. (McTaggart 1927: §§ 305, 306, 310; cf. also McTaggart 1908: 458-59)

What is said to hold for events occupying, or being located at, B-series positions, is also meant to hold for 'moments of absolute time', i.e. time-instants, 'if such moments should exist' (McTaggart 1927: §310).

Thus, according to McTaggart, for every two distinct B-series positions *x* and *y*, either *x* precedes *y* or *y* precedes *x*, while for every two events *x* and *y*, either *x* and *y* are contemporaneous, in that there is a B-series position at which they are co-located, or else either *x* precedes *y* or *y* precedes *x*. Although he does not mention this explicitly, we may also say that, according to McTaggart, any two moments of time are contemporaneous iff they are identical, and if they are distinct, one precedes the other. For McTaggart, then, the B-series is a total ordering of B-series positions and moments of time, i.e. time-instants. Secondly, for every *x*, *y* and *z*, if *x* precedes *y* and *y* precedes *z*, then *x* precedes *z*, while for every *x* and *y*, if *x* precedes *y*, *y* does not precede *x*. Consequently, no *x* precedes itself. Precedence is transitive, asymmetric and hence also irreflexive.

If we presume, as McTaggart would seem to do, that occupants of B-series positions only ever occupy one such position, and so are instantaneous, we may unproblematically add to this, not only that any such occupant is contemporaneous with itself and that for any two such occupants, *x* and *y*, *x* is contemporaneous with *y* iff *y* is contemporaneous with *x*, but furthermore that for any three such occupants *x*, *y* and *z*, if *x* is contemporaneous with *y*, and *y* with *z*, then *x* is also contemporaneous with *z*. For such instantaneous occupants of B-series positions, contemporaneity accordingly is reflexive, symmetrical and transitive. As indicated, contemporaneity is best understood in terms of co-location at some B-series position, where time-instants are co-located iff identical. In fact, once time-instants enter the picture, we can discard quantification over B-series positions in favour of quantification over time-instants, and conceive of location as location at time-instants, as long as we allow time-instances to be located at themselves.

So far, so good. However, McTaggart makes a further claim, *viz.* that B-series relations are *permanent*, and he concludes from this that the occupants of B-series positions can never come into existence or go out of existence. Following Broad (1938: 290, 296, 298–300), we can reconstruct McTaggart's reasoning as follows. The sense in which McTaggart claims B-series relations to be permanent is best captured by the following principle, where 'REL' may uniformly be substituted for by either 'precedes', 'is located at' or 'is contemporaneous with'

(1) Always, $\forall x$ Always, $\forall y$ Always, (x REL $y \rightarrow$ Always, x REL y)

(cf. also Sider 2001: 12). Thus, in line with what we said about identity in the previous chapter, (1) takes REL to be *eternal*. From this McTaggart goes on to infer the following conclusion which delivers the intended result that things that are ever REL-related, never come into or go out of existence:

(2) Always, $\forall x$ Always, $\forall y$ Always, (x REL $y \rightarrow$ Always, E!x & E!y)

The validity of the inference from (1) to (2), however, essentially depends on

(3) Always, $\forall x$ Always, $\forall y$ Always, (x REL $y \rightarrow$ E!x & E!y)

In other words, it depends on the assumption that REL is *existence-entailing* and so obeys Williamson's *being constraint*. But once this assumption is granted, McTaggart's characterisation of the permanence of REL, i.e. (1), will – quite unsurprisingly – prove unacceptable to temporaryists, because temporaryists *qua* temporaryists deny

(4) Always, $\forall x$ Always, $\forall y$ Always, (E!x & E!$y \rightarrow$ Always, (E!x & E!y))

Instead, temporaryists will then characterise the permanence of REL in terms of its *rigidity* as follows:

(5) Always, $\forall x$ Always, $\forall y$ Always, (x REL $y \rightarrow$ Always, ((E!x & E!y) $\rightarrow x$ REL y))

If REL is *eternal* in the sense of (1), it is *rigid* in the sense of (5), but the converse implication does not hold; and it is now clear that once (5) takes the place of (1), (2) can no longer be derived. Alternatively, temporaryists may endorse (1), but reject the presupposition that REL is existence-entailing, i.e. (3), thereby once again blocking the inference to (2).

McTaggart himself ultimately offered an analysis of precedence in tensed terms that allows us to reject (2). According to McTaggart, x precedes y just in case sometimes, x is present and y is future, or alternatively, just in case sometimes, x is past and y is present (McTaggart 1927: §610; cf. also Prior 1967: 4; 2003: 143). On this analysis, precedence is not existence-entailing and so fails to underwrite (3), and yet is eternal in the sense of (1), since it is a theorem of the tense logic introduced in Chap. 1 that what sometimes is the case always, sometimes is the case. We might give corresponding analyses of location and contemporaneity, with similar results, by saying that *x* is located at *y* just in case sometimes, *y* is a time-instant and *x* and *y* are both present, and that *x* is contemporaneous with *y* just in case sometimes, both *x* and *y* are present.

It is to Broad's credit that he clearly saw, and theoretically exploited, that B-series relations are not the exclusive domain of permanentist ontologies (*pace* McTaggart and also Sider 2001: 12–14), but can just as well play an important role in the articulation of temporaryist ontologies, once either (1) or (3) is rejected. Broad himself opted for acceptance of (3) at the cost of rejecting (1) (Broad 1923: 66–67; see however page 64, where Broad briefly invokes McTaggart's definition of precedence). But it serves the temporaryist's purposes equally well to accept (1) while rejecting (3). In fact, if REL is assumed to be existence-entailing but not eternal, we can define a corresponding relation that is eternal but not existence-entailing in terms of 'Sometimes, *x* REL *y*'; and if REL is assumed to be eternal but not existence-entailing, we can define a corresponding relation that is existence-entailing but not eternal in terms of 'E!*x* & E!*y* & *x* REL *y*'. Either approach has the additional benefit that it becomes ultimately quite unnecessary to appeal to monadic properties of being present, past or future (see Chaps. 4 and 5; cf. also Correia and Rosenkranz 2015b).[1]

However, arguably, if contemporaneity is existence-entailing, it will not even be rigid in the sense of (5), once we give up on the idea that occupants of B-series positions can at most ever occupy one such position (i.e. can at most ever be located at one time-instant). Thus, consider two distinct pulsations, e.g. two distinct discontinuous series of periodically recurring discharges of energy with different frequencies, that have never been, but now are synchronized. Insofar as we count such pulsations themselves as objects in time, we may have reason to say that they are now co-located, but even if they already existed sometime in the past, had never been co-located before, to the extent that now did not exist before. If co-location fails to be rigid, it is hard to see how location and precedence could be.

In what follows, we will construe all B-series relations as eternal but not existence-entailing.

3.2 Temporal Location

McTaggart speaks of positions in the B-series standing to each other in relations of precedence, and contrasts these with time-instants, i.e. 'moments of absolute time' that, if they existed, would occupy such positions and be identical iff they occupied the same such position. He is sceptical, however, whether there are any such time-instants. Positions are distinct from their occupants; and for all that has been said,

[1] In later work, however, Broad explicitly endorsed the contravening thought that statements involving tense-inflections and adverbial modifiers could be reduced to statements involving temporal adjectives like 'present', 'past' and 'future' (as well as metricised versions of the latter two) together with a tensed copula (Broad 1938: 271–73).

there could be unoccupied B-series positions. McTaggart's B-series positions might therefore be taken to be similar to abstract times, or to the kinds of 'times of a model' that we invoke when doing formal semantics.

By contrast, we will quantify over times (time-instants) which we take to be whenever they exist concrete rather than abstract, and to stand in relations of precedence, and to correspond to occupied B-series positions in McTaggart's sense. We will not quantify over B-series positions in addition. We leave it open whether times are *sui generis* entities or are identical to, say, the sum total of events occupying the corresponding B-series position. We likewise leave it open whether times are sets of facts that we would otherwise describe as obtaining at those times. We will speak of things other than times as being contemporaneous with a given time just in case they are co-located at that time. With these preliminaries being in place, we can now proceed to lay down principles governing the notion of location at times in whose terms the notion of contemporaneity, or co-location, can be defined.

Something can only ever be located at a time; and time x is located at time y iff x is identical to y. We let 'L' be short for 'is located at', and accordingly lay down:

(A25) $x \mathrm{L} y \rightarrow \mathrm{T}y$
(A26) $\mathrm{T}x \rightarrow (x \mathrm{L} y \leftrightarrow x = y)$

We furthermore take location to be eternal – in other words, we assume

(A27) $x \mathrm{L} y \rightarrow$ Always, $x \mathrm{L} y$

By contrast, we do *not* assume location to be existence-entailing.

Not everything is ever located at some time: abstract things like numbers are not. Only things *in time* meet this condition. To be in time *just is* sometimes to be located at some time:

(D6) m is in time $\equiv_{\mathrm{df}}$ Sometimes, $\exists x(m \mathrm{L} x)$

Accordingly, being in time is not existence-entailing.

It follows from (A26), (D6) and (A16), i.e. 'Sometimes, E!m', that times are themselves in time. We next define what it is to be a *resident of time* as follows:

(D7) R$m \equiv_{\mathrm{df}} m$ is in time & Always, $\forall x$ Always, ($m \mathrm{L} x \rightarrow$ (E!$x \rightarrow$ E!m)) & Always, (E!$m \rightarrow \exists y(m \mathrm{L} y)$)

Times and events are residents of time. Yet, not all things in time are residents of time. Plausibly, if x is a set that has a given resident of time y as a member, x is located wherever y is located. Thus, if x is {noon, midnight}, x is located at noon and so a thing in time. But insofar as at noon, midnight did not yet exist, neither did x, all the while, by (A23), i.e. 'T$x \rightarrow$ At x, E!x', at noon, noon already existed. Accordingly, x fails to be a resident of time. Just as sets are located where their members are located, fusions are located where their parts are located. Yet, while a set exists only when all of its members exist, a mereological fusion, by contrast, exists whenever one of its parts exists. So, the fusion of the number seven and Fabrice is in time, but given temporaryism, it may exist even before any time exists at which Fabrice is located; and since the number seven is never located at any time, the fusion may accordingly already have existed long before it ever was located. Therefore, the fusion of the number seven and Fabrice is in time, but not a resident of time.

Although we are mostly interested in residents of time, since quantification is assumed to be unrestricted, the contrast between things and things in time, and the contrast between things in time and residents of time, do matter.

We next define being *instantaneous* as follows:

(D8) m is instantaneous $\equiv_{\text{df}}$ m is in time & Always, $\forall x$ Always, $\forall y(m \text{ L } x$ & $m \text{ L } y \rightarrow x = y)$

It follows, as desired, that moments of time, and events of zero duration, are instantaneous. But it also follows that both the fusion of the number seven and, say, a given flash of lightning, and the set of the number seven and that flash of lightning, are instantaneous.

We can now also define contemporaneity in terms of co-location at times as follows:

(D9) $m \approx n \equiv_{\text{df}}$ Sometimes, $\exists x(m \text{ L } x$ & $n \text{ L } x)$

Just like being in time, contemporaneity is not existence-entailing. Given (D6) and (D9), it follows both that whatever is in time is contemporaneous with itself and that contemporaneity is symmetric:

$$x \text{ is in time} \rightarrow x \approx x$$
$$x \approx y \rightarrow y \approx x$$

Accordingly, contemporaneity is symmetric and is reflexive for things in time. Since contemporaneous times are identical, and since instantaneous things in time quite generally are only ever located at one time, for such things in time, contemporaneity is also transitive. For non-instantaneous things in time, by contrast, contemporaneity cannot be assumed to be transitive: given (D9), Edgar Allan Poe and Johann Wolfgang von Goethe were contemporaries, and so were Poe and Queen Victoria, but Goethe had already died when Queen Victoria was born.

This concludes our formal characterisation of location and contemporaneity. The axioms and definitions allow for more substantive characterisations of what these notions involve. Relativistic physics tells us that location and contemporaneity are only ever relative to a foliation of spacetime into spacelike hypersurfaces, and that so are times. The principles laid down here are in principle suited to permit such further relativisation: given any such foliation, they specify how location and contemporaneity behave according to that foliation. We will return to this issue in Chaps. 8 and 9.

3.3 Precedence

The relation of precedence (being earlier than) and the relation of succession (being later than) are interdefinable: x precedes y just in case y succeeds x. Everything we want to say by means of the latter, we can just as well say by means of the former. With hindsight it is better to first characterise a relation of precedence for times and to define the general notion applicable to all things in time, including times, in its terms. We symbolise the restricted relation by '$\prec$' and lay down:

(A28) $x \prec y \rightarrow (\mathrm{T}x \,\&\, \mathrm{T}y)$

We take this relation to be *eternal* (but not to be existence-entailing), thus:

(A29) $x \prec y \rightarrow \text{Always}, (x \prec y)$

We also lay down:

(A30) $\neg(x \prec x)$
(A31) $(x \prec y \,\&\, y \prec z) \rightarrow x \prec z$
(A32) $(\mathrm{T}x \,\&\, \mathrm{T}y) \rightarrow (x \prec y \vee x = y \vee y \prec x)$

(A30) and (A31) respectively say that the relation $\prec$ is irreflexive and that it is transitive. That the relation be transitive is an essential ingredient of our most basic conception of the temporal order; that it be irreflexive, by contrast, is a substantive metaphysical assumption: it rules out circular time, for instance. As we shall see in more detail later, unlike its rivals, the Growing Block Theory of time in fact requires that precedence be irreflexive, since it implies that what exists constantly grows as time goes by. However, making this assumption here does not in turn disadvantage any of the other contenders: all varieties of permanentism and temporaryism alike can without problems accept it. From (A30) and (A31) it follows that the relation $\prec$ is also asymmetric. (A32) says that $\prec$ is a total ordering on the set of times.

We now define the general notion of precedence, which we symbolise by '$<$':

(D10) $m < n \equiv_{\mathrm{df}} n$ is in time & Sometimes, $\exists x(m\,\mathrm{L}\,x \,\&\, \text{Always}, \forall y(n\,\mathrm{L}\,y \rightarrow x \prec y))$

In other words, m precedes n just in case both are things in time and m is sometimes located at a time that precedes any time at which n is ever located. Given (D10), (A25) and (A26), always, any two times stand in the relation $<$ if, and only if, they stand in the relation $\prec$. Given (D10), the axioms (A30) and (A31) yield the corresponding principles as theorems:

(T7) $\neg(x < x)$
(T8) $(x < y \,\&\, y < z) \rightarrow x < z$
(T9) $x < y \rightarrow \neg(y < x)$

Accordingly, like $\prec$, the relation $<$ is irreflexive, transitive and asymmetric. Notice, however, that unlike '$x \prec y$' and '$x \approx y$', '$x < y$' and '$x \approx y$' do not mutually exclude each other: although, by (D9), Queen Victoria and Edgar Allan Poe were contemporaries, by (D10), Poe also precedes Queen Victoria, because Poe had been located at a time preceding any time at which Queen Victoria was located.

The following mixed axioms, which will prove crucial later, mimic the usual truth-clauses for H and G:

(A33) $(\text{At}\, x, \text{H}\varphi) \leftrightarrow (\text{T}x \,\&\, \text{Always}, \forall y(y \prec x \rightarrow \text{At}\, y, \varphi))$
(A34) $(\text{At}\, x, \text{G}\varphi) \leftrightarrow (\text{T}x \,\&\, \text{Always}, \forall y(x \prec y \rightarrow \text{At}\, y, \varphi))$

where y is not free in φ and where x and y are distinct variables. What, at a given time t, has always been the case, is whenever any time earlier than t exists, the case at that earlier time; and what, at a given time t, is always going to be the case, is whenever any time later than t exists, the case at that later time. These left-to-right conditionals across (A33) and (A34) should impress as eminently plausible. Their converses are not as obvious, but on reflection equally plausible: what is, whenever any time earlier than t exists, the case at that earlier time, has, at t, always been the case; and what is, whenever any time later than t exists, the case at that later time, is, at t, always going to be the case. Let us give an informal proof of the right-to-left conditional across (A33); corresponding considerations would show the right-to-left conditional across (A34). Assume that whenever a time t' earlier than t exists, at that earlier time t', φ is the case. Assume, contrary to (A33), that nevertheless at t, sometimes in the past, $\neg\varphi$ was the case. What is always the case, always has always been the case, and what always holds, holds at t in particular. So, at t, always in the past, for any time t' earlier than t, at t', φ. Very plausibly, however, always in the past, anything that is the case, is the case at what then is the present time, and *vice versa*. Equally plausibly, at t, always in the past, any time that is present precedes t, and hence does so eternally. Accordingly, both (i) sometimes in the past, there exists a time earlier than t at which $\neg\varphi$, and (ii) always in the past, for any time t' earlier than t, at t', φ. Evidently, (i) and (ii) cannot both hold at t.

This concludes our formal characterisation of precedence. The axioms and definitions allow for more substantive characterisations of what the notion involves. Relativistic physics tells us that precedence is only ever relative to a foliation of spacetime into spacelike hypersurfaces, and that so are times. The principles laid down here are in principle suited to permit such further relativisation: given any such foliation, they specify how precedence behaves according to that foliation. We will return to this issue in Chaps. 8 and 9.

We have now assembled all the conceptual and logical tools needed to properly articulate the characteristic tenets of GBT and to derive what follows from them. This task will be undertaken in the next chapter. In Chaps. 8 and 9, we will address the problems that relativistic physics holds in store for GBT and, in effect, all other temporaryist ontologies. Until then, we will presume a non-relativistic background physics. As we shall argue later, this way of proceeding will prove beneficial once we try to come to terms with these problems and revise GBT in their light.

Chapter 4
The Growing Block

Abstract In this chapter we reconstruct the original version of the Growing Block Theory of time first advanced by C. D. Broad, highlight its shortcomings, and propose an improved version of the theory. We show that this improved version of the theory is superior to two more recent attempts to capture the idea of the growing block. In Sect. 4.1 we critically review central passages from Broad's *Scientific Thought*, identify core principles that give substance to the image of a growing block, delimited by an edge of becoming beyond which nothing exists, and diagnose a number of problems with Broad's account. In Sect. 4.2 we then present a neater version of the theory that still incorporates central ideas of Broad's, yet avoids those problems. In Sect 4.3 we critically review the accounts respectively advanced by M. Tooley and T. Button and conclude that our version of the theory fares much better.

In this chapter we begin by reconstructing, and critically discussing, the original version of GBT first advanced by C. D. Broad in his 1923 *Scientific Thought* – a task which will occupy us throughout Sect. 4.1. Broad's text is admirably clear and allows us to identify a number of principles, some that we deem essential ingredients of GBT, others that are inherently problematical and should therefore be abandoned. However, it turns out that even when purified of the latter, Broad's theory, though coherent, remains in an important sense incomplete: while it ensures that there is a growing block of being with an edge of becoming beyond which nothing exists, it is as yet silent on where, at any given moment, this edge of becoming lies. Accordingly, in Sect. 4.2, we suggest an alternative characterisation of GBT, consisting of just two principles, that at once fills this lacuna and allows us to recover all the tenable principles of Broad's original view. The result is a simplified and powerful version of GBT whose formulation requires only minimal resources, *viz.* logical constants, temporal operators, quantification, identity and the notion of being a time-instant. As we argue in Sect. 4.3, this makes it far superior to other, more recent attempts at formulating GBT which, without exception, deploy unfamiliar conceptual machinery. Our version of GBT shows that no such unfamiliar conceptual machinery is needed to capture the idea of the growing block, and thereby renders itself immune to powerful objections levelled against those attempts.

F. Correia, S. Rosenkranz, *Nothing To Come*, Synthese Library 395,
https://doi.org/10.1007/978-3-319-78704-6_4

4.1 The Broad Picture

4.1.1 Constant Growth

We owe the idea of reality as a growing block to C. D. Broad who characterises it as a combination of at least two thoughts:

> The sum total of existence is always increasing [...]. (Broad 1923: 66-67)

> There is no such thing as *ceasing* to exist; what has become exists henceforth for ever. (Broad 1923: 69)

Plausibly, Broad is not here concerned with things in time such as the odd fusions or sets alluded to in Sect. 3.2 above, but rather exclusively with residents of time. Accordingly, for Broad, always there is a new resident of time that was nothing before, while always everything will always in the future be something. Thus:

(B1) $\mathrm{E!}x \rightarrow \mathrm{GE!}x$
(B2) $\exists x(\mathrm{R}x \;\&\; x \text{ is new})$

where, to recall,

(D7) $\mathrm{R}m \equiv_{\mathrm{df}} m$ is in time & Always, $\forall x$ Always, $(m \mathrm{~L~} x \rightarrow (\mathrm{E!}x \rightarrow \mathrm{E!}m))$ & Always, $(\mathrm{E!}m \rightarrow \exists y(m \mathrm{~L~} y))$

and where *being new* is defined as follows:

(D11) m is new $\equiv_{\mathrm{df}} \mathrm{E!}m \;\&\; \mathrm{H}\neg\mathrm{E!}m$

Broad also ventures to say that '[w]hatever is has become' (Broad 1923: 69). This suggests:

$$\mathrm{E!}x \rightarrow (x \text{ is new} \vee \mathrm{P}(x \text{ is new}))$$

But the latter principle is far too strong. Broad himself adds:

> For complete accuracy a slight modification ought to be made in the statement that 'whatever is has become'. Long events do not become [...] as wholes. Thus the becoming of a long event is just the successive becoming of its shorter sections. (Broad 1923: 69)

However, as long as such 'long events' do not extend into the infinite past, they satisfy the aforementioned principle – even if, when they are new, they will have some new sections in the future – and so, in this sense, 'do not become as wholes'. Accordingly, the modification Broad suggests is not necessary. But it is also insufficient to make his statement that 'whatever is has become' any more acceptable, precisely because there may after all be residents of time that have always existed in the past, or that, although they haven't always existed in the past, always in the past, sometimes in the past already existed. There is no evident reason why GBT should be at odds with admission of such things.

Both (B1) and (B2) would seem essential to the idea of temporal reality as a growing block. For, (B1) alone merely says that nothing will cease to exist, which is as yet consistent with always everything always being something, and so, contrary to what is intended, with the *absence of becoming*. (B2) alone merely says that always some resident of time is new, which is as yet consistent with always some such thing being lost, and so contrary to what is intended, with the block's *erosion*. Only taken together do these principles imply that always the domain of what there is properly includes the domain of what there once was, which is precisely what the image of the growing block is meant to capture. Broad writes:

> The sum total of existence is always increasing, and it is this which gives the time-series a sense as well as an order. A moment t is later than a moment t' if the sum total of existence at t includes the sum total of existence at t' together with something more. (Broad 1923: 66-67)

As already highlighted in Chap. 3, there is textual evidence that Broad takes temporal relations, such as precedence and location, to be existence-entailing:

> When an event, which was present, becomes past, it does not change or lose any of the relations which it had before; it simply acquires in addition new relations which it *could* not have before, because the terms to which it now has these relations were then simply non-entities. (Broad 1923: 66)

Although we resolved to deploy notions of precedence and location that are *not* existence-entailing, the latter can be used to define corresponding notions that *are* existence-entailing.[1] Bearing this in mind, we can safely articulate what Broad here says about precedence among moments of time in the following terms:

(B3) $x \prec y \leftrightarrow (\text{At}\, x, \forall z\, \text{At}\, y, \text{E!}z) \mathbin{\&} (\text{At}\, y, \exists z\, \text{At}\, x, \neg\text{E!}z)$

Above we suggested that (B1) and (B2) jointly imply that always the domain of what there is properly includes the domain of what there once was. Indeed, (B3) can be shown to follow from (B1) and (B2) together with neutral principles:

Proof: (i) Assume $x \prec y$. Next assume At x, E!z. From this by (B1), we get: At x, GE!z. From this, by a theorem we obtain from (A34), *viz.*

(T10) $x \prec y \rightarrow ((\text{At}\, x, \text{G}\varphi) \rightarrow \text{At}\, y, \varphi)$ (with y not free in φ)

we get: At y, E!z. Accordingly, we have: $x \prec y \rightarrow (\text{At}\, x, \text{E!}z \rightarrow \text{At}\, y, \text{E!}z)$, and so: $x \prec y \rightarrow \text{At}\, x, (\text{E!}z \rightarrow \text{At}\, y, \text{E!}z)$. But then: Always, $\forall z(x \prec y \rightarrow \text{At}\, x, (\text{E!}z \rightarrow \text{At}\, y, \text{E!}z))$. Given that $x \prec y$, Always, $\forall z(x \prec y)$, and so we get: Always, $\forall z$ At x, (E!$z \rightarrow$ At y,

[1] The definitions are

m PREC $n \equiv_{\text{df}}$ E!m & E!n & $m < n$
m LOC $n \equiv_{\text{df}}$ E!m & E!n & m L n

E!z)). By (A24) we conclude: At x, $\forall z$ At y, E!z. (ii) Assume again $x \prec y$. From (B2), we then have: At y, $\exists z$H$\neg$E!z. Hence by (A24): Sometimes, $\exists z$ At y, (E!z & H$\neg$E!z). By a theorem we obtain from (A33), *viz.*

(T11) $x \prec y \rightarrow$ ((At y, Hφ) $\rightarrow$ At x, φ) (with x not free in φ)

we then get: Sometimes, $\exists z$ (At y, E!z & At x, $\neg$E!z), and so: Sometimes, $\exists z$ At y, (E!z & At x, $\neg$E!z). By (A24) again, we conclude: At y, $\exists z$ At x, $\neg$E!z. (iii) Assume (At x, $\forall z$ At y, E!z) & (At y, $\exists z$ At x, $\neg$E!z). Then Tx & Ty and $x \neq y$. So by the totality axiom (A32), $x \prec y$ or $y \prec x$. Assume $y \prec x$. By the result obtained under (ii), this contradicts the initial assumption. Hence, $x \prec y$. □

Broad goes on to characterise precedence among events as follows:

> [W]hen we say that the red section [of the history of a signal lamp] precedes the green section, we mean that there was a moment when the sum total of existence included the red event and did not include the green one, and that there was another moment at which the sum total of existence included all that was included at the first moment and also the green event. (Broad 1923: 67)

Note that, by (A16), (B1) anyway entails

Sometimes, (E!x & E!y)

Accordingly, the passage last quoted suggests that, for Broad, one event precedes another just in case sometimes, the former exists without the latter, i.e.

(B4″) x and y are events $\rightarrow$ ($x < y \leftrightarrow$ Sometimes, (E!x & $\neg$E!y))

What holds for events in general, *a fortiori* holds for momentary events. Momentary events are *instantaneous* residents of time ordered by precedence, in the sense of the earlier definition:

(D8) m is instantaneous $\equiv_{\text{df}}$ m is in time & Always, $\forall x$ Always,$\forall y$(m L x & m L $y \rightarrow x = y$)

As residents of time, momentary events only ever exist when there is a time at which they are located, and are guaranteed to exist whenever a time exists at which they are located. Given that momentary events are also instantaneous, if, in general, one momentary event x precedes another y just in case sometimes x exists while y does not exist, then the same will hold for times. Against the background of these reflections, we can here take Broad to subscribe to the following principle:

(B4′) Tx & Ty $\rightarrow$ ($x \prec y \leftrightarrow$ Sometimes, (E!x & $\neg$E!y))

Extended events, i.e. events that need time to unfold, can be conceived of as *aggregative fusions* of momentary events. Unlike other fusions, extended events are themselves residents of time in the sense of (D7). A natural extension of (B4″) accordingly is

(B4) $Rx \,\&\, Ry \rightarrow (x < y \leftrightarrow \text{Sometimes}, (E!x \,\&\, \neg E!y))$

That Broad intends this more general principle is plausible. After all, he subscribes to a very liberal conception of events according to which residents of time such as the cliffs at Dover count as events:

> By an *event* I am going to mean anything that endures at all, no matter how long it lasts or whether it be qualitatively alike or qualitatively different at adjacent stages in its history. (Broad 1923: 54)

Since (B4) in turn entails (B4′), the ensuing discussion will focus on the more general principle, (B4).

Broad writes as if (B4) was a consequence of (B3); and in the presence of (B3), it is indeed implausible to think of (B4) as an entirely separate axiom. That is to say, as principles governing the notion of precedence, (B3) and (B4) should have a common root. However, (B1) and (B2) do not jointly entail (B4).[2] As we shall see in the next section, (B3) and (B4) are both derivable from a set of principles that likewise allow us to derive (B1) and (B2).

4.1.2 *The Edge of Becoming*

That there be a single edge of becoming beyond which nothing exists is as much part of the picture of the growing block as are (B1) and (B2). However, bearing in mind the distinction between existence and location, it would seem that, so far, we cannot exclude that some of the new things in time are located at different times. The distinction itself should be uncontroversial. As Craig Bourne explains,

> [we can] distinguish *existing* at a place/time, from being *located* at a place/time. Just as I can say that Socrates exists as of this time (if I believe in the real existence of the past), I can say that Australia exists as of this place (given I don't think that England is the only real place). But this is, of course, not to say that Socrates is located now, or that Australia is located here [i.e. in England]. Thus, I maintain, it does make sense to talk about, and is perfectly natural to talk about, objects existing at times and places other than those at which they are located […]. (Bourne 2006: 164-65)

[2] To see this, note that while the conjunction of (B1) and (B2) is compatible with the idea that times permanently exist, (B4) is not. Even if we replaced (B2) by the stronger '$\exists x(Tx \,\&\, x$ is new)', (B4) could not be derived. Thus, imagine that time is discrete and has a beginning, that the first two times t_1 and t_2 are both new at t_1, and that at t_2, t_3 is new, at t_3, t_4 is new, and so on. Let this combination of claims be (V). (V) entails both '$\exists x(Tx \,\&\, x$ is new)' and the negation of (the universal closure of) (B4). Accordingly, if the conjunction of (V) and (B1) entailed (B4), this would mean that this conjunction is inconsistent. But, on the face of it, the conjunction of (V) and (B1) is *not* inconsistent.

Broad offers further principles whose truth would ensure that there is a single edge of becoming beyond which nothing exists. Thus, he writes:

> Let us call [the change from future to present] *Becoming*. [W]hen an event becomes, it *comes into existence*; and it was not anything at all until it had become. You cannot say that a future event is one that succeeds the present; for a present event is defined as one that is succeeded by nothing. (Broad 1923: 67-68; emphases in the original)
>
> [T]he essence of a present event is, not that it precedes future events, but that there is quite literally *nothing* to which it has the relation of precedence. (Broad 1923: 66; emphasis in the original)

According to Broad, then, present events have just come to exist, having never existed before, and being present, do not precede anything. Broad here speaks of events in general. If what he says holds for events in general, it *a fortiori* holds for momentary events. Similarly, since momentary events are instantaneous residents of time, in the sense defined earlier, what he says holds for momentary events iff it holds for the times at which they are located. Against the backdrop of these reflections, we can take the passages just quoted to suggest the following two principles:

(B5) $\mathrm{T}x \rightarrow (x \text{ is present} \leftrightarrow x \text{ is new})$
(B6) $\mathrm{T}x \rightarrow (x \text{ is present} \leftrightarrow x \text{ is last})$

where being last can be defined as follows:

(D12) $m \text{ is last} \equiv_{\mathrm{df}} \mathrm{E}!m \;\&\; \neg\exists x(\mathrm{R}x \;\&\; m < x)$

(B5) and (B6) already ensure that momentary events are present iff new iff last. However, (B5) does not lend support to the more general claim that *all* events, momentary or extended, are present iff new; and (B6) does not lend support to the more general claim that all events are present iff last. To see this, consider Émilien's life and the life of his little sister Agathe: Émilien's life is present, but already existed in the past and, by (D10), precedes the life of Agathe. Charitably interpreted, however, and given what he says elsewhere about the utility of extensive abstraction for the formulation of theoretical principles, Broad is here perhaps best understood to be exclusively concerned with momentary events (Broad 1923: 54-57).

(B5) and (B6) jointly entail

(B7) $\mathrm{T}x \rightarrow (x \text{ is new} \leftrightarrow x \text{ is last})$

In fact, (B2) and (B4) independently allow us to establish the more general claim, entailing (B7), that always every resident of time is new whenever last, and last whenever new, i.e.

(B8) $\mathrm{R}x \rightarrow (x \text{ is new} \leftrightarrow x \text{ is last})$

Proof: (i) Assume that Rm, and that m is new. Assume for *reductio* that there is a resident of time n such that m precedes n. By (B4), sometimes in the past, m exists while n does not. But this contradicts the assumption that m is new. (ii) Assume that Rm, and that m is last. Assume for *reductio* that PE!m. Given (B2), presently some resident of time n is new. From this and (B4) it follows that some resident of time succeeds m, which contradicts the assumption that m is last. □

From (B7), in conjunction with the totality axiom (A32), we can derive:

(B9) (Tx & x is new & Ty & y is new) → $x = y$

Proof: Assume Tx & Ty, and: x is new & y is new. Assume for *reductio*: $x \neq y$. Then by (A32), either $x < y$ or $y < x$. But given (B7), *ex hypothesi* neither option is possible. □

In other words, then, (B7) and (A32) allow us to derive that there is at most one new time. It follows that any two new residents of time are co-located at the sole time that is new. Accordingly, and contrary to first appearances, already by appeal to (B2) and (B4) alone, Broad can ensure that there is a single edge of becoming beyond which nothing exists.

4.1.3 Where Does the Edge of Becoming Lie?

However, (B1), (B2) and (B4) alone do not yet imply that the uniquely new time is new *at itself* and hence neither that new residents of time are located at the very time at which they are new and not, say, at a later time. In other words, while (B1), (B2) and (B4) are sufficient to determine that there is a single edge of becoming beyond which nothing exists, they are insufficient to determine its whereabouts at any given moment.

It is at this juncture that principles (B5) and (B6) – interderivable in the light of (B7) – would seem to reassert their significance. For, as long as it can be presupposed as trivial that

(#) Tx → At x, x is present

holds, either of those principles would ensure that every time is new at itself. As we have seen, however, in laying down a principle like (B6), Broad professes to *define* what it is for an instantaneous event to be present, in the sense of disclosing its *essence* qua being present (Broad 1923: 66-68). He even ventures to say that

> the predicates, past, present, and future, are of their very nature relational [...]. (Broad 1923: 65)

But then, (#) should *follow* from a characterisation of what it is to be last, together with (B6), rather than having to be invoked on independent grounds. For, unless it can be taken for granted that any time is last at itself, it remains an open question whether the notion of presentness whose nature (B6) is meant to reveal is the very same notion of presentness underwriting (#).

It is of course in line with the project of revealing the essence, or nature, of a given property that we have an independent, prior conception of that property, encoded in a number of principles that strike us as trivial; and it is likewise evident that, in laying down (B6), Broad is not in the business of providing a stipulative definition of 'is present' in application to times (or by extension, momentary events), but rather intends to give a partial real definition of the property that our common notion of being present denotes. Plausibly, this common notion underwrites (#) which would seem part of our independent, prior conception of presentness. However, according to Broad, what is last will in the future cease to be last; and so if (B6) reveals the nature of being present, at least as applied to times (and by extension to momentary events), presentness must be a property that can be lost. Arguably, though, our independent, prior conception of presentness as a property that can be lost, if such there be, is of a property that is *non-relational*. At least, this much is what most dynamic theories of time, which put presentness to theoretical use, would seem to take for granted (see Chap. 5). This common conception may, of course, ultimately be misguided in more than one respect; but then this merely goes to show that Broad cannot simply cherry-pick those parts of our ordinary conception whose truth would serve him to discharge his theoretical obligations, here: to determine where, at any given moment, the edge of becoming lies. These reflections enforce the general point that in order to be acceptable, the *definiens* of Broad's (partial) real definition (B6) must independently be shown to fulfil all those principles of presentness on which his version of GBT ultimately relies. Yet, the conjunction of (B1), (B2) and (B4) fails to deliver the desired result that every time is new, and so last, at itself.

As we have argued, there is another loose end. While (B3) could ultimately be derived from (B1) and (B2), the connection between (B3) and (B4) remained somewhat obscure. Yet, both principles govern the notion of precedence and so should not be independent.

4.2 GBT Reloaded

Our discussion so far suggests that all of the following principles are essential ingredients of GBT, and so should be validated by any version that improves upon Broad's own:

(B1) $E!x \rightarrow GE!x$
(B2) $\exists x(Rx \,\&\, x \text{ is new})$
(B3) $x \prec y \leftrightarrow (\text{At } x, \forall z \text{ At } y, E!z) \,\&\, (\text{At } y, \exists z \text{ At } x, \neg E!z)$

We saw that we could prove (B3) from the conjunction of (B1) and (B2). Thus, (B1) and (B2) adequately capture the idea of a block that always grows without ever eroding.

Similarly, any improvement upon Broad's conception of the growing block should vindicate the following:

(B7) $\mathrm{T}x \rightarrow (x \text{ is new} \leftrightarrow x \text{ is last})$
(B9) $(\mathrm{T}x \,\&\, x \text{ is new} \,\&\, \mathrm{T}y \,\&\, y \text{ is new}) \rightarrow x = y$

(B7) and (B9) encode the equally central idea that there is a single edge of becoming beyond which nothing exists. As we saw, (B9) could be derived from (B7) with the help of (A32). Given other principles Broad accepted, (B7) itself was derivable in two different ways. The first of these ways relied on Broad's account of presentness, embodied by the two principles

(B5) $\mathrm{T}x \rightarrow (x \text{ is present} \leftrightarrow x \text{ is new})$
(B6) $\mathrm{T}x \rightarrow (x \text{ is present} \leftrightarrow x \text{ is last})$

The derivation of (B7) is immediate. However, we saw that this choice embroiled us in controversies concerning the ordinary notion of presentness that these principles invoke. Such concerns about the notion of presentness become further aggravated in the light of recent criticisms, to be reviewed in Chap. 5, which suggest that this notion is ill-suited for heavy-duty use in the context of metaphysical theorizing. So, with hindsight, we had better set aside (B5) and (B6). The second route by which Broad could establish (B7) relied on (B2) and

(B4) $\mathrm{R}x \,\&\, \mathrm{R}y \rightarrow (x < y \leftrightarrow \text{Sometimes, } (\mathrm{E!}x \,\&\, \neg \mathrm{E!}y))$

which allowed demonstration of a principle that is more general than (B7), *viz.*

(B8) $\mathrm{R}x \rightarrow (x \text{ is last} \leftrightarrow x \text{ is new})$

However, (B4) seemed to serve no other purpose and seemed anyway strangely disconnected from (B3), the other principle governing precedence that Broad invoked. If we can derive (B7) – and, perhaps, (B8) – by other means, there is no need to invoke (B4). If we can derive both (B3) and (B4) on independent grounds, thereby showing that they have a common root, so much the better.

The task that lies ahead of us accordingly is to improve upon Broad's version of GBT while vindicating at least (B1), (B2) and (B7). We now suggest adopting Broad's principle (B1) as our first axiom, relabelling it '(P1)':

(P1) $\mathrm{E!}x \rightarrow \mathrm{GE!}x$

We argued that Broad's conception of the growing block, while sufficient to determine that there is a single edge of becoming, is insufficient to determine its whereabouts. In particular, it failed to validate the central thought that every time is new at itself, so that any resident of time freshly added to the block is *located at the time of its addition*. Instead of trying to derive this thought by other means, we explicitly lay down, as the second of our axioms for GBT,

(P2) $\quad \mathrm{T}x \rightarrow \mathrm{At}\, x, \mathrm{H}\neg\mathrm{E!}x$

We now argue that (P1) and (P2), taken together, are sufficient to characterise GBT. To begin with, we show that (P2) entails (B2):

Proof: Recall one of the axioms governing the At-operators introduced in Chap. 3:

(A22) $\quad \forall x(\mathrm{T}x \rightarrow \mathrm{At}\, x, \varphi) \rightarrow \varphi$ (with x not free in φ)

From (P2), (A23), according to which any time exists at itself, and (A9), according to which any time is always a time, we get

$$\forall x(\mathrm{T}x \rightarrow \mathrm{At}\, x, \exists y(\mathrm{T}y \,\&\, y \text{ is new}))$$

and so, we can derive (B2) using (A22). □

We saw in the previous section that (B1) – *aka* (P1) – and (B2) jointly entail (B3). Accordingly, the same result can be obtained from (P1) and (P2). We thus have vindicated all Broadian principles that were meant to capture the idea of the block as constantly growing without ever eroding.

This leaves the task of vindicating (B7) – and thereby also (B9) – because it is ultimately this principle that ensures that there is a single edge of becoming beyond which nothing exists. As it turns out, (B7) is readily derivable from (P1) and (P2).

Proof: (i) Assume Tm & E!m & H¬E!m. Assume for *reductio* E!n & Tn & $m < n$. From (T11), we have: $x \prec y \rightarrow ((\mathrm{At}\, y, \mathrm{H}\neg\mathrm{E!}y) \rightarrow \mathrm{At}\, x, \neg\mathrm{E!}y)$. From (P2), we have: At n, H¬E!n. From the foregoing by (A23), we get: At m, (E!m & ¬E!n). By (T2), this yields: Sometimes, E!m & ¬E!n. Given the definition of 'Sometimes' and (P1), the latter contradicts the initial assumptions. We have thus established: Tm & m is new → ¬∃x(Tx & $m < x$). Let's now establish: Tm & m is new → ¬∃x(Rx & $m < x$). Assume Tm & m is new. Assume for *reductio* E!n & Rn & $m < n$. Then there is a time n^* such that n L n^*. By the definition of <, it follows that $m < n^*$. But by the previously established result, this is impossible. (ii) Assume Tm & E!m and ¬∃x($m < x$). Assume for *reductio*, PE!m. From (P2), (A23) and (A22) we have: ∃y(Ty & y is new). Let n be a witness so that Tn & n is new. *Ex hypothesi*, both $n \neq m$ and ¬($m < n$). So by the totality axiom (A32), $n < m$. By the same reasoning as in (i), with the roles of m and n reversed, we get a contradiction. □

With (B7) being in place, we can now derive (B9) in the way set out in the previous section:

(B9) (Tx & x is new & Ty & y is new) $\to x = y$

Accordingly, (P1) and (P2) are also jointly sufficient to capture the idea of an edge of becoming beyond which nothing exists.

(P2), (A23) and (B7) yield:

(T12) T$x \to$ At x, x is new
(T13) T$x \to$ At x, x is last

Using (T2) and (T5) one can get:

(T14) (Sometimes, $\exists x$ At x, φ) $\leftrightarrow$ Sometimes, φ (with x not free in φ)

We now use (T14) and (T13), alongside (P1) and (P2), to prove Broad's

(B4) Rx & R$y \to$ ($x < y \leftrightarrow$ Sometimes, (E!x & ¬E!y))

Proof: (i) Suppose Rx & Ry and $x < y$. Then by (D10), Sometimes, $\exists z$(x L z & Always, $\forall z'$ (y L $z' \to z \prec z'$)). But then, given (A27) and (A12), Sometimes, $\exists z$ At z, (x L z & $\forall z'$ (y L $z' \to z \prec z'$)). By (A23), (T13) and (D7), it follows that Sometimes, $\exists z$ At z, (E!x & ¬E!y), and hence by (T14) that Sometimes, (E!x & ¬E!y). (ii) Suppose Rx & Ry and E!x & ¬E!y. By (D7), there is a time, z, such that x L z. Suppose for *reductio* that Sometimes, $\exists z'$ (y L z' & ¬ ($z \prec z'$)). Note first that the following is a theorem:

(α) $z' \prec z \to$ Always, (E!$z \to$ E!z')

Suppose indeed that $z' \prec z$ but Sometimes, (E!z & ¬E!z'). By (P1), then, Always, (E!$z' \to$ E!z). By (A23), (A12) and (A29), we infer that At z', z' is not last, which contradicts (T13). Using (α), it is then easy to establish that the following is a theorem:

(β) ($z' \prec z$ & Ry & y L z') $\to$ Always, (E!$z \to$ E!y)

Let us now establish that the initial hypotheses lead to a contradiction. One of the hypotheses is that Sometimes, $\exists z'$ (y L z' & ¬ ($z \prec z'$)). Given that by hypothesis, E!z and Ry, we then have thanks to (A32) that Sometimes, $\exists z'$ (y L z' & $z' \prec z$)). But given (β), we then have that Sometimes, Always, (E!$z \to$ E!y), and hence that Always, (E!$z \to$ E!y). It then follows that E!y – contrary to assumption. Hence, we have established that (Rx & Ry & E!x & ¬E!y) $\to \exists z$(x L z & Always, $\forall z'$ (y L $z' \to z \prec z'$)) is a theorem. But then Rx & R$y \to$ (Sometimes, (E!x & ¬E!y) $\to$ Sometimes, $\exists z$(x L z & Always, $\forall z'$ (y L $z' \to z \prec z'$)) is also a theorem. By (D10), then, Rx & R$y \to$ (Sometimes, (E!x & ¬E!y) $\to x < y$) is a theorem. □

We now likewise have

(B8) $\mathrm{R}x \rightarrow (x \text{ is new} \leftrightarrow x \text{ is last})$

Given (B9) and (T12), as well as

(T5) $\varphi \rightarrow \exists x\, \mathrm{At}\, x, \varphi$ (with x not free in φ)
(A8) $(\mathrm{At}\, x, \varphi) \rightarrow \mathrm{T}x$
(A9) $\mathrm{T}x \rightarrow \mathrm{Always}, \mathrm{T}x$
(A12) $(\mathrm{Always}, \varphi) \rightarrow (\mathrm{T}x \rightarrow \mathrm{At}\, x, \varphi)$
(A13) $(\mathrm{At}\, x, \varphi) \rightarrow \mathrm{Always}, \mathrm{At}\, x, \varphi$

we can in addition prove the following two theorems:

(T15) $(\mathrm{T}x \,\&\, x \text{ is new}) \rightarrow (\varphi \leftrightarrow \mathrm{At}\, x, \varphi)$

Proof: Suppose that Tm & m is new & φ. Then by (T5), ∃x At x, (Tm & m is new & φ), with x distinct from m and not occurring free in φ. By (A8), (A9), (T12), it follows that ∃x At x, (Tm & Tx & m is new & x is new & φ). By (B9) it follows that At m, φ. We have just established that '(Tm & m is new) → (φ → At m, φ)' is a theorem. It follows that '(Tm & m is new) → (¬At m, ¬φ → φ)' is also a theorem. But it follows from this that '(Tm & m is new) → ((At m, φ) → φ)' is also a theorem. □

(T16) $(\mathrm{At}\, x, \varphi) \leftrightarrow (\mathrm{T}x \,\&\, \mathrm{Always}, (x \text{ is new} \rightarrow \varphi))$

Proof: The left-to-right direction across the biconditional in (T16) already follows from (T15), (A8), (A13) and basic principles of tense logic. The right-to-left direction follows from (A12) and (T12). □

In other words, then, if a time is new, it is also *accurate*: something holds at that time just in case it holds (for the notion of accuracy, see Dorr and Goodman forthcoming). Moreover, if something holds at a given time, it holds whenever that time is new, and *vice versa*. These are two important results which show that, in application to times, *being new* fulfils the function traditionally assigned to *being present*, when it comes to elucidating how operators of the form 'At m' work. Note, however, that we do not take these observations to suggest that 'is present' should be defined in terms of 'is new'.

Accordingly, given (P1) and (P2), and the neutral principles introduced in Chaps. 1, 2 and 3, we can establish (B1) to (B4), and (B7) to (B9), and hence all of the Broadian principles worth saving. In particular, they show (B3) and (B4) to share a common root. At the same time, (P2), which says that every time is new at itself, guarantees that any newly added resident of time is located at the time of its addition. (P1) and (P2) therefore not only guarantee that the block of being is constantly growing without ever eroding, and that it is delimited by a single edge of becoming beyond which nothing exists, they furthermore determine where, at any given moment, this edge of becoming lies. In addition, (P1) and (P2) entail that the new time is accurate, and that whatever holds at a time, holds whenever that time is new, and *vice versa*. So, while our version of GBT clearly improves, and fruitfully expands, upon Broad's original, it remains most faithful to it.

All the while, (P1) and (P2) only require the leanest ideology: they only deploy logical constants, temporal operators, quantification, identity and the notion of being a time. In particular, *no* notion of being present needs to be invoked anywhere along the way. In their sum, we take these results to provide overwhelming *prima facie* reason to regard the combination of (P1) and (P2) as the most adequate, powerful and elegant formulation of GBT. Whether this is so in the end depends on how the proposed characterisation of GBT fares in comparison with other extant versions of the view. Besides Broad's own, there are two alternative accounts that dominate the contemporary discussion of the growing block, the one proposed by Michael Tooley (1997) and the one proposed by Tim Button (2006 and 2007). However, before comparing our version of GBT to these competitors in the next section, we close this section by examining which of the Barcan formulas, and converse Barcan formulas, proponents of (P1) and (P2) ought to accept, and which ones they ought to reject.

Consider the following schematic formulae, with 'Ω' being a placeholder for a particular operator and φ an arbitrary formula:

(BF_Ω) $\Omega\exists x\varphi \rightarrow \exists x\Omega\varphi$
(CBF_Ω) $\exists x\Omega\varphi \rightarrow \Omega\exists x\varphi$

(BF_Ω) is the *Barcan formula* for 'Ω', while (CBF_Ω) is the *converse Barcan formula* for Ω. We can accordingly ask, given a particular choice of 'Ω', whether it validates either formula. Answers to these questions cannot always be given in theory-neutral terms.

Given (P1), according to which always everything will always in the future be something, it seems plain that GBT allows for (BF_Ω) to be valid once 'Ω' is replaced by 'P'. Recall

(A1) $\varphi \rightarrow HF\varphi$
(A2) $\varphi \rightarrow GP\varphi$
(A3) $H(\varphi \rightarrow \psi) \rightarrow (H\varphi \rightarrow H\psi)$
(A4) $G(\varphi \rightarrow \psi) \rightarrow (G\varphi \rightarrow G\psi)$
(A15) $\forall x E!x$
(A17) $\forall x(\varphi \rightarrow \psi) \rightarrow (\forall x\varphi \rightarrow \forall x\psi)$
(A18) $\varphi \rightarrow \forall x\varphi$ (x not free in φ)
(A19) $(\forall x\varphi \,\&\, E!m) \rightarrow \varphi[m/x]$
(R1) $\varphi \,/\, H\varphi$
(R2) $\varphi \,/\, G\varphi$
(R3) $\varphi \,/\, \forall x\varphi$

Using these axioms and rules and (P1), we can indeed derive

(BF$_P$) $P\exists x\varphi \rightarrow \exists xP\varphi$

Proof: By (A19), $E!x$ & $P\varphi \rightarrow \exists xP\varphi$. By (R2) and (A4), then, $GE!x$ & $GP\varphi \rightarrow G\exists xP\varphi$. Thanks to (P1) and (A2), then, $E!x$ & $\varphi \rightarrow G\exists xP\varphi$. Using (A15), (A17), (A18) and (R3), we then get $\exists x\varphi \rightarrow G\exists xP\varphi$. By application of the rule $\varphi \rightarrow G\psi$ / $P\varphi \rightarrow \psi$, which we get from (A1), (A3) and (R1), we derive $P\exists x\varphi \rightarrow \exists xP\varphi$. □

It seems equally plain that, given (P1), (CB$_\Omega$) is valid once 'Ω' is replaced by 'F' or 'G' – in other words, that according to GBT, both of the following formulae are valid:

(CBF$_F$) $\exists xF\varphi \rightarrow F\exists x\varphi$
(CBF$_G$) $\exists xG\varphi \rightarrow G\exists x\varphi$

This is confirmed by the following two proofs that, in addition to (P1), make use of

(A4) $G(\varphi \rightarrow \psi) \rightarrow (G\varphi \rightarrow G\psi)$
(A15) $\forall xE!x$
(A17) $\forall x(\varphi \rightarrow \psi) \rightarrow (\forall x\varphi \rightarrow \forall x\psi)$
(A18) $\varphi \rightarrow \forall x\varphi$ (x not free in φ)
(A19) $(\forall x\varphi$ & $E!m) \rightarrow \varphi[m/x]$
(R2) φ / $G\varphi$
(R3) φ / $\forall x\varphi$

Proof: By (A19), $E!x$ & $\varphi \rightarrow \exists x\varphi$. By (R2) and (A4), then, $GE!x$ & $F\varphi \rightarrow F\exists x\varphi$. Thanks to (P1), then, $E!x$ & $F\varphi \rightarrow F\exists x\varphi$. Using (A15), (A17), (A18) and (R3), we then get $\exists xF\varphi \rightarrow F\exists x\varphi$. □

Proof: By (A19), $E!x$ & $\varphi \rightarrow \exists x\varphi$. By (R2) and (A4), then, $GE!x$ & $G\varphi \rightarrow G\exists x\varphi$. Thanks to (P1), then, $E!x$ & $G\varphi \rightarrow G\exists x\varphi$. Using (A15), (A17), (A18) and (R3), we then get $\exists xG\varphi \rightarrow G\exists x\varphi$. □

Unlike the case of (BF$_P$), the Barcan formula for 'H' fails: Always in the past, 'now' referred to some time, but (unless time is beginning now) it is not the case that there is something that always in the past, was referred to by 'now'. The same open formula "now' refers to x' can be employed to discredit (BF$_G$) and (BF$_{Always}$).

(B2) can be used to show that (BF$_F$) fails. Suppose for *reductio* that it holds. By the derived rule '$F\varphi \rightarrow \psi$ / $\varphi \rightarrow H\psi$', we can derive '$\exists x\varphi \rightarrow H\exists xF\varphi$'. Let φ be 'x is new'. By (B2), the antecedent is true. But (unless time is beginning now), the consequent is false. (B2) also suffices to argue against (CBF$_\Omega$) with 'Ω' replaced by 'P', 'H' or 'Sometimes'. Thus, let φ be 'x is nothing'. Then (unless now is the first moment of time) by (B2), for these choices of 'Ω' the antecedent of (CBF$_\Omega$) is true, but since never anything is nothing, the consequent of (CBF$_\Omega$) fails. Incidentally, this falsifies Prior's contention that on any view that admits 'past existents but not future ones', (CBF_P) holds (Prior 1967: 171).

Since, according to temporaryism, 'Sometimes, $\exists x$(x is a house John built)' may be true, while '$\exists x$ Sometimes, (x is a house John built)' is not, (BF_Ω) likewise proves invalid once 'Ω' is replaced by 'Sometimes'. Similarly, if we let φ be 'E!$x \rightarrow x$ = Fabrice's first son', (CBF_Ω) will prove invalid once 'Ω' is replaced by 'Always'. Thus, while '$\exists x$ Always, (E!$x \rightarrow x$ = Fabrice's first son)' is true, 'Always, $\exists x$(E!$x \rightarrow$ x = Fabrice's first son)' is not.

What about (BF_Ω) and (CBF_Ω) once 'Ω' is replaced by operators of the form 'At m'? Note that there are at least as many such temporal operators as there are terms for times. Let x be some time before Émilien's birth. Then according to GBT, '$\exists y$ At x, y is nothing' is true, while 'At x, $\exists y$(y is nothing)' is not. So (CBF_{At}) is not generally valid. The same is true of the following *quasi*-converse Barcan formula

($CBF_{At})^{\forall}$ $\forall x((\exists y \text{ At } x, \varphi) \rightarrow \text{At } x, \exists y\varphi)$ (with x and y distinct variables)

Next consider:

(BF_{At}) $(\text{At } m, \exists y\varphi) \rightarrow \exists y \text{ At } m, \varphi$

where y is distinct from m. (BF_{At}) fails. If it held, then by (P2), 'T$m \rightarrow \exists y$ At m, H¬E!y' would be a theorem. But let m be the referent of 'now'. If 'T$m \rightarrow \exists y$ At m, H¬E!y' were a theorem, then '$\exists y$ At m, H¬E!y' would be always true, and so would be true one day ago. But this would imply that one day ago, there existed something which did not exist, which is impossible.

Now consider the following *quasi*-Barcan formula:

($BF_{At})^{\forall}$ $\forall x((\text{At } x, \exists y\varphi) \rightarrow \exists y \text{ At } x, \varphi)$,

where x and y are distinct variables. ($BF_{At})^{\forall}$ turns out to be provable, given

(T2)	$(\text{At } x, \varphi) \rightarrow \text{Sometimes}, \varphi$
(T12)	$Tx \rightarrow \text{At } x, x$ is new
(T16)	$(\text{At } x, \varphi) \leftrightarrow (Tx \;\&\; \text{Always}, (x \text{ is new} \rightarrow \varphi))$
(BF_P)	$P\exists x\varphi \rightarrow \exists xP\varphi$

Proof: Using (T2) and (T12), we derive 'E!$x \rightarrow$ ((At x, $\exists y\varphi$) $\rightarrow$ Sometimes, (x is new & $\exists y\varphi$))'. Given our basic tense logic and the definition of 'x is new', 'E!$x \rightarrow$ ¬F(x is new)' is a theorem, and so we can derive 'E!$x \rightarrow$ ((At x, $\exists y\varphi$) $\rightarrow$ ((x is new & $\exists y\varphi$) ∨ P(x is new & $\exists y\varphi$)))'. Now since x and y are distinct variables, '(x is new & $\exists y\varphi$) ∨ P(x is new & $\exists y\varphi$)' entails '$\exists y$(x is new & φ) ∨ P$\exists y$(x is new & φ)', which, by (BF_P), in turn entails '$\exists y$(x is new & φ) ∨ $\exists y$P(x is new & φ)', and hence '$\exists y$((x is new & φ) ∨ P(x is new & φ))'. Since the latter yields '$\exists y$ Sometimes, (x is new & φ)', we can derive 'E!$x \rightarrow$ ((At x, $\exists y\varphi$) $\rightarrow \exists y$ Sometimes, (x is new & φ))'. Using (T16) we can then derive 'E!$x \rightarrow$ ((At x, $\exists y\varphi$) $\rightarrow \exists y$ At x, φ)', which entails ($BF_{At})^{\forall}$.

□

Table 4.1 Barcan and converse Barcan: GBT

	Barcan Formula	**Converse Barcan Formula**
	GBT	
P	Yes	No
H	No	No
F	No	Yes
G	No	Yes
Sometimes	No	No
Always	No	No
At *m*	No	No
	$(\mathbf{BF}_{\mathbf{At}})^{\forall}$	$(\mathbf{CBF}_{\mathbf{At}})^{\forall}$
	GBT	
	Yes	No

So, given the temporal operators in play, (BF_P) and $(BF_{At})^{\forall}$ are the only relevant (*quasi-*)Barcan formulas, and (CBF_F) and (CBF_G) are the only relevant converse Barcan formulas that we can take to be valid according to GBT. Table 4.1 summarises these results.

We have thus managed to put GBT on a simple foundation, consisting of just two, very powerful axioms, thereby reducing the number of independent principles:

(P1) $E!x \rightarrow GE!x$
(P2) $Tx \rightarrow At\ x, H\neg E!x$

(P1) and (P2) are themselves very simple and deploy only familiar and well-understood notions. So, although in this section and the previous one, we had to go through many proofs – some of them straightforward, others more tedious – in order to see the inferential power of these axioms unfold, the resultant theory is extremely easy to grasp. As we shall see in the next chapter, GBT's main competitors – presentism and permanentism, both in its dynamic and static varieties – allow for equally simple characterisations and can moreover be conceived in similarly familiar terms. In particular, and contrary to common lore, neither of these other contenders needs to invoke the notion of presentness, which, as we shall soon be able to more fully appreciate, proves an important asset. But before we turn to a discussion of these other views, we will, in the remainder of this chapter, critically examine two recent attempts to cash out the idea of the growing block. Unlike our version of GBT, these attempts are fraught with difficulties, which further confirms that our version is the growing blocker's best shot.

4.3 No Funny Business

Now that our own preferred version of GBT is on the table, it is worthwhile briefly comparing it to other theories, besides Broad's own, that have been advertised under the same label. A quick glance at (P1) and (P2) reveals that, besides familiar logical connectives, our version of the view only uses quantification, temporal operators, identity and the notion of being a time. The ways these notions behave have painstakingly been set out in previous chapters. The ideology of our version of GBT is accordingly rather sparse and imports no unfamiliar or obscure notions. So one dimension of comparison should be whether other contemporary versions of GBT draw on an equally sparse set of clearly defined concepts.

Our version of GBT is a species of temporaryism which employs quantification that is both unrestricted and tensed. As such, it can accommodate a key thought of Broad's original proposal, *viz.* that what exists changes with time, yet without construing it as involving restricted quantification over a domain of objects that only permanentists can avail themselves of. So another dimension of comparison is whether other contemporary versions of GBT can do equal justice to the thought that 'the sum total of existence' at one time differs from the 'the sum total of existence' at any other time, without helping themselves to an ontology of things of which, at all times, both these sums are mere subsets. At the very least, all versions of GBT should be dynamic rather than static. Vindicating temporaryism is one way to achieve this goal. So an intimately related dimension of comparison is whether other extant versions of GBT are equally well-equipped to secure that the universe is dynamic rather than static.

These are clearly not the only relevant dimensions of comparison. Ultimately, the comparative assessment will also have to relate to each version's capacity, or incapacity, to successfully answer extant philosophical challenges – the most pressing of which we will address in Chaps. 6, 7, 8 and 9. However, we will here confine our attention to the three dimensions just stated. Moreover, we will focus on just two competitors: Michael Tooley's version of GBT, elaborated in his 1997 book *Time, Tense and Causation*, and Tim Button's *no-futurism*, expounded in a series of articles (Button 2006 and 2007). While Button's more recent work has sparked a small debate with Jonathan Tallant (Tallant 2007 and 2011), almost all of the recent commentators on GBT refer to Tooley (1997) rather than Broad (1923) as the key text (see e.g. Miller 2013).

According to Tooley, what exists *as of* one time, differs from what exists *as of* another (Tooley 1997: 16). In particular, on Tooley's version of GBT, if time x is later than time x', what exists *as of* x includes, but comprises more than, what exists *as of* x'. This seems to simply echo Broad's claim that 'a moment t is later than a moment t' if the sum total of existence at t includes the sum total of existence at t' together with something more' (Broad 1923: 66-67; cf. Tooley 1997: 173). However, as we shall see in due course, this interpretation would be mistaken.

Tooley thinks that if a theory has the consequence that what exists *as of* one time differs from what exists *as of* another, this alone is quite sufficient to ensure that the theory is dynamic rather than static. In particular, he argues that it need not, and ultimately should not, be assumed *in addition* that quantification and predication must be tensed, if we are to capture such cross-temporal differences in what there is. Tooley (1997: 19-20) writes:

> Consider [...] the view of time according to which, while the past and present are real, the future is not, and suppose that such a view is true of our own world. Then the states of affairs that are actual as of the year 1990 do not include any that involve purple sheep, whereas, given appropriate advances in genetic engineering, the states of affairs that are actual as of the year 2000 might very well do so. But such a difference is one that, on the face of it, can be described without using any tensed terms, since it is simply a matter of there being a spatiotemporal region in which various non-temporal properties, such as that of being purple, are instantiated, and which is actual as of the year 2000, but not as of the year 1990. The assumption that tenseless temporal concepts are semantically basic appears to be perfectly compatible, therefore, with the possibility that the world is a dynamic one. [...] The metaphysical hypothesis that the world is a static one does entail that there are no irreducible tensed facts, and therefore that tensed concepts cannot be semantically basic. But, on the other hand, the hypothesis that the world is a dynamic one does not entail that tenseless temporal concepts cannot be semantically basic. [...] This means, in turn, that a dynamic world need not involve any special, irreducible tensed properties [...] in order for tensed sentences to be true: it may simply be a world where what tenseless states of affairs are actual is different at different times.

On Tooley's own version of the view, quantification is indeed taken to be tenseless, and existence is accordingly taken to be timeless (Tooley 1997: 40-41, 127, 149, 155-56, 188). Truths *simpliciter* about what timelessly exists capture 'the totality of existence' (Tooley 1997: 155). Moreover, according to Tooley, all positive statements can be brought into existential form, quantifying over states of affairs or events, and can be given tenseless truth conditions (Tooley 1997: 191-204). Granted this and assuming the equivalence of 'As of x, $\neg\exists x\varphi$' and 'Tx & $\neg$(As of x, $\exists x\varphi$)', we only have to concern ourselves with the significance of statements of the form 'As of x, $\exists x\varphi$', where '$\exists$' is tenseless.

The characterisation so far has two important consequences. First, operators of the form 'As of x' crucially differ from temporal operators of the form 'At x'. Secondly, in the light of this finding, operators of the form 'As of x', lest they remain obscure, emerge as simple means to restrict tenseless quantification in ways that make the resultant theory vulnerable to hostile takeover by eternalists, i.e. proponents of a static universe.

To see this, first reflect that as far as temporal operators of the form 'At x' go, for any tenseless φ, both the inference from φ to 'At x, φ' and its converse are valid. For, if φ is tenseless, prefixing it with a temporal operator can never effect a change in truth-value. In particular then, if x is later than x', and if the clause embedded in the truth 'As of x, $\exists y(y = x)$' is tenseless, as Tooley claims, the assumption that operators of the form 'As of x' work in exactly the same way as temporal operators of the form 'At x', would have the unwanted implication that 'As of x', $\exists y(y = x)$' is likewise true.[3]

[3] Tooley uses 'It is true at x that' interchangeably with 'As of x', which is a bit confusing since the former sports an 'at' where one would expect an 'as of', and so is more suggestive of a reading according to which it is equivalent to 'At x'.

Accordingly, on the one hand, the idea that what exists as of one time differs from what exists as of any earlier time, invites a reading of 'As of *x*' as being equivalent in meaning to 'At *x*'. But on the other hand, Tooley's insistence on tenseless quantification precludes that very reading.

Let us accordingly ask how else statements of the form 'As of *x*, $\exists y\varphi$' are to be understood on Tooley's version of GBT. Tooley himself does not provide any alternative interpretation, but rather contents himself with treating the locution '*x* exists as of *y*' as primitive (Tooley 1997: 40-41). Of course, not all theoretical notions need to be defined, but in the light of the foregoing, more must be said here in order to defuse the worry that there might simply be no coherent notion at all that could fit the bill.

Consider a statement of the form 'As of *x*, $\exists y\varphi$' where φ is tenseless. For all ends and purposes, we can here assume that the quantifier exclusively ranges over residents of time. Then given that, according to Tooley, quantification is also tenseless and only ranges over timelessly existing things, we might lay down the following equivalence:

$$(\text{As of } x, \exists y\varphi) \leftrightarrow \exists y(\varphi \,\&\, \exists x' \,((x' \prec x \vee x' = x) \,\&\, y \,\mathrm{L}\, x'\,))$$

where '$\exists$' is again presumed to be tenseless. If φ is tensed, matters are slightly more complicated, but it is a fair guess that we can elucidate the intended reading of 'As of *x*, $\exists y\varphi$' in such cases by saying that, again *modulo* GBT on Tooley's construal of it, the following equivalence holds:

$$(\text{As of } x, \exists y\varphi) \leftrightarrow \exists y((\text{At } x, \varphi) \,\&\, \exists x' \,((x' \prec x \vee x' = x) \,\&\, y \,\mathrm{L}\, x'\,))$$

where '$\exists$' is again tenseless. In both cases, 'As of *x*, $\neg\exists y\varphi$' and 'T*x* & $\neg$(As of *x*, $\exists x\varphi$)' may be treated as being equivalent.

It now immediately follows that what exists as of today includes, but comprises more than, what exists as of *x*, whenever *x* is earlier than today. It likewise follows that 'As of today, $\exists y\varphi$' does not entail 'As of yesterday, $\exists y\varphi$'. However, for all that has been said, operators of the form 'As of *x*' now simply emerge as handy ways of singling out segments of a static four-dimensional manifold over whose inhabitants we can, by Tooley's own admission, always quantify (tenselessly and *simpliciter*). In other words, it becomes entirely unclear why Tooley's account cannot straightforwardly be appropriated by proponents of a static conception of time, since according to his account, no tensed concepts are basic, and the operators he employs merely restrict the otherwise unrestricted quantifier which latter, according to Tooley, has a permanent domain.[4]

Tooley himself nonetheless insists that his theory cannot be appropriated by defenders of a static view (Tooley 1997: 200-201, 213). But when he insists on the impossibility of such hostile takeover, he would seem to equivocate between (i) the

[4] See Oaklander 2004: 137-39, for a similar diagnosis; and see Tooley 1997: 19, for what seems like an unwitting endorsement of the very idea.

claim that eternalists cannot allow that what, at x, exists may differ from what, at x', exists, and (ii) the distinct claim that they cannot allow that what, *as of* x, exists may differ from what, *as of* x', exists (Tooley 1997: 201). According to both his own theory and eternalism, what, at x, exists cannot differ from what, at x', exists, since existence is meant to be timeless. But, equally, according to either view, what both exists *and is located at x or some time preceding x* may differ from what exists *and is located at some time preceding x*.

Accordingly, either Tooley's view invokes an obscure notion whose task description would seem to impose mutually incompatible demands, or else it fails to forestall hostile takeover by proponents of a static universe, thereby betraying one of the core ideas of GBT.

Articulating the position that he calls 'no-futurism', Tim Button (2006 and 2007) too invokes an unfamiliar locution, reminiscent of Tooley's, *viz.* 'x is real-as-of y'. Yet Button is more careful in this respect and happy to accept the ineliminability of tense. Indeed, much of the letter of what Button says would seem congenial to the version of GBT which we set out in this chapter, because in central places his use of 'x is real-as-of y' invites recapture in terms of 'At y, E!x', where the embedded clause is tensed. However, appearances are again misleading. For, Button (2007: 331) also writes:

> No-futurists claim that, as-of any moment, only earlier and simultaneous moments are real. Someone might ask: is the 'are' in the thesis to be read as tensed or tenseless? This question misunderstands no-futurism. The 'is' in 'x is real-as-of y' is neither tensed nor tenseless. The relation 'x is real-as-of y' is a primitive of no-futurism, and the verb it includes is just part of that primitive. Tense [...] supervene[s] upon real-as-of relations, and tenseless truths [...] follow in their wake. But it is clearly therefore nonsensical to demand that the relation itself must be either tensed or tenseless.

As Tallant (2011) has argued, this conception of 'x is real-as-of y' as a relational expression that is neither tensed nor tenseless, raises more questions than it answers. For one thing, postulating expressions that belong to some third category besides those respectively subsuming tensed and tenseless expressions, not only seems entirely *ad hoc*, it would result in a language altogether different from any language with which we are familiar, regimented or not. If the only way 'no-futurism' can coherently be formulated is in terms of such an unfamiliar language – and Button seems quite adamant that such is the case – this would rather count against it.

One mistaken idea, underlying Button's proposal, would seem to be that, even after careful analysis, 'x is real-as-of y' is a relational expression. If, instead, we construe it in tense-logical terms as 'At y, E!x' – and Button himself frequently uses 'as-of y' as an operator rather than as part of a relational expression (e.g. Button 2006: 132 and 2007: 330) – then the perceived pressure to say that it is neither tensed nor tenseless evaporates. 'At y, E!x' clearly belongs to the regimented language of tense logic, since it is composed of a temporal operator and an embedded clause in the present tense. It nonetheless has a stable truth-value, given that, to recall, the following is an axiom of the tense logic we devised in Chap. 1:

(A13) $(\text{At } x, \varphi) \rightarrow \text{Always, At } x, \varphi$

Alternatively, if one thinks, for whatever reason, that only those sentences qualify as tensed that do not have a stable truth-value, then one should be equally willing to classify sentences of the form 'At *y*, E!*x*' as tenseless. Either way, there is no temptation to think that such sentences belong to some third category. Yet, equally, either way, as long as we construe 'At *y*' as a temporal operator that shifts the circumstances of evaluation of the embedded clause to those prevailing at *y*, the embedded clause is in the present tense. Why would this be at odds with no-futurism? Button does not say.

Taking the embedded clause to be in the present tense certainly cannot be said to be incompatible with the idea, central to no-futurism, that '*x* is real-as-of *y*' behaves asymmetrically. For surely, 'At *y*, E!*x*' does not logically entail 'At *x*, E!*y*'. Perhaps the thought is that if 'Martian outposts exists' was in the present tense, the truth of 'In the year 3000, Martian outposts exist' would imply the present existence of Martian outposts. But this would evidently be to misconstrue the way temporal operators function. Or perhaps the thought is that if the clause 'dinosaurs exist' was in the present tense, then endorsing 'At the present time, dinosaurs exist' would commit us to the claim that dinosaurs are located at the present time. That is for example the kind of thought that Broad himself succumbed to in later years, when he distanced himself from his 1923 view, saying that 'the metaphor of the history of the world "growing continually longer in duration by the addition of new slices", which I took seriously in *Scientific Thought*' is misleading insofar as it 'presupposes that phases, which have already supervened and been superseded, in some sense "co-exist" with each other and with that which is now happening' (Broad 1959: 767; cf. also Broad 1938: 307). However, as long as we systematically distinguish between the co-existence of two residents of time *a* and *b*, expressible by '$\exists x \exists y(a = x \mathbin{\&} b = y)$', and their co-location (or contemporaneity), expressible by '$\exists x(a \mathrm{~L~} x \mathbin{\&} b \mathrm{~L~} x)$', it is altogether unclear why the metaphor should be misleading (cf. Bourne 2006: 164-65).

All in all, Button's reasons for insisting that no-futurism requires the locution '*x* is real-as-of *y*' as a primitive that is neither tensed nor tenseless remain obscure. The version of GBT set out in this chapter anyway avoids postulation of such a primitive relation and hence all the problems such postulation brings in its wake. As we shall argue in Chap. 6, it also successfully undermines Tallant's claim that Button's account is the only 'no future' account that holds out the promise of avoiding recent sceptical challenges (Tallant 2011: 44-45).

The idea that defenders of GBT must resort to unheard-of conceptual resources to stabilize their view is a persistent theme in the extant literature on the topic. Here is Ted Sider on the matter:

> Broad and Tooley want to say that a current utterance of 'it once was the case that the entire four-dimensional reality contained only one world war' is true, since in 1935, for example, the growing block universe only contained what had occurred up until that point. However, if we evaluate the component sentence 'the entire four-dimensional reality contains only one world war' with respect to 1935 (let me stipulate that 'the entire four-dimensional reality' is to apply to *all* of reality), we obtain falsehood. The reason is that the component sentence concerns all of reality rather than just the 'time of evaluation', and hence evaluat-

ing the sentence with respect to 1935 is the same as evaluating the sentence for truth *simpliciter*. Since reality (now) contains a second world war, the sentence is false. A similar point can be made by invoking the notion of 'the crest of the wave', which is the present edge of reality, the portion of reality such that no event exists after it. The crest of the wave is, while I write this sentence, in 2000, but, [proponents of the growing block theory] want to say, it once was 1935. The problem is that the proposed analysis of a current utterance of 'WAS-64-years-ago (the crest of the wave is present)' seems false, since when we inspect the 1935 slice of reality we find no crest.

These examples show that the defender of the growing block universe must accept two senses of the tenses. One sense is given an eternalist-style analysis in terms of the manifold; the other captures the growth in the manifold. [...] On the first sense, the tenses are in an important sense relative to times, since we need a reference point – the time of the token of a tensed sentence – to give an evaluation for truth. The tenses on the second reading are not relative in this way: it is true simpliciter that reality used to be smaller, and will be larger. (Sider 2001: 22)

As can be double-checked by consulting the principles we identified as characteristic of GBT, formulation of the theory requires only one 'sense of tense' – i.e. the one at work in standard tense logic with its temporal operators and its ultimately present-tensed atoms. So something must be wrong with Sider's diagnosis.

To see what is wrong with it, let us first compare the different behaviour of the terms 'now' and 'the present time'. The term 'now' is an indexical that always 'takes wide scope' over whatever temporal operators embed the clause in which it occurs. The term 'now' always refers to the time of utterance; and here it does not matter how deep inside the scope of temporal operators the token of the indexical 'now' is buried. By contrast, on at least one pertinent understanding of it, the phrase 'the present time' (just like the phrase 'the referent of 'now'') does not always refer to the time of utterance, and so does not 'take wide scope' over whatever temporal operators embed the clause in which it occurs. On this understanding, any temporal operator prefixing the simple present-tensed clause in which the phrase 'the present time' occurs shifts its reference to the time to which this temporal operator shifts the evaluation. To illustrate, the following two sentences are true at the time of writing:

Eight years ago, now is a time later than any time in 2017
Eight years ago, the present time is not later than any time in 2017

Note that, in order to draw the contrast between the ways in which these two terms respectively behave, we do not need to invoke two different types of tense. Some phrases have their reference always determined by the time of utterance, while others have their reference determined by the relevant time of evaluation; and while sometimes the time of utterance is the relevant time of evaluation – e.g. when the phrase is used in an unembedded present-tensed clause – sometimes the time of utterance is not the relevant time of evaluation – e.g. in certain cases when the phrase is used in a clause embedded by a time-of-evaluation-shifting temporal operator.

Once this is taken to heart, it is clear that, for the proponent of GBT, the phrases 'the edge of becoming', and 'the entire fourdimensional reality', will behave like the phrase 'the present time', and emphatically not like the indexical 'now'. Accordingly, on the growing block view, when we semantically 'evaluate the component sentence "the entire four-dimensional reality contains only one world war" with respect to 1935', we do *not* obtain falsehood, precisely because, as Sider notes, in 1935, 'the growing block universe only contained what had occurred up until that point'; and this is so irrespective of the fact that when we *now* use that sentence, the phrase 'the entire fourdimensional reality' denotes a reality including WW2, which would render our present use of that sentence false.

This is unsurprising. After having walked exactly one mile, it may be true to say that the part of the way we've covered then is a third of the way to go, while after having walked yet another mile, it is no longer true to say that the part of the way we've covered then is a third of the way to go.

Similarly, and bearing in mind that the temporal operator 'Back in 1935' operates on a present-tensed clause, on the growing block view, a current utterance of 'Back in 1935, the edge of becoming is present' is true, because back in 1935, the edge of becoming is located at some time in 1935. Sider's contention that this 'seems false, since when we inspect the 1935 slice of reality we find no crest', is evidently based on a false assumption. Thus, he would seem to presuppose that, given only that the edge of becoming now is located elsewhere – namely, at some time later than any time in 1935 – the occurrence of the phrase 'the edge of becoming', being embedded by the operator 'Back in 1935', refers to some time later than any time in 1935. But this just means that Sider treats the phrase 'the edge of becoming' as if it functioned like the indexical 'now' – which is evidently wrong. (Of course, one may define a phrase like 'the-edge-of-becoming-as-it-is-now'; and then it will be true to say, in 2017, that, even back in 1935, the-edge-of-becoming-as-it-is-now is located at some time in 2017 and not in 1935. However, the proponent of GBT need have no qualms with that.)

In light of these diagnoses, our version of GBT emerges as clearly superior. The foregoing sections have already demonstrated that, despite its sparse ideology, this version of GBT is quite resourceful. However, GBT is commonly taken to face a number of formidable challenges. In Chaps. 6, 7, 8 and 9, we will discuss three of the best known such challenges, the so-called epistemic objection, the challenge to account for the truth of future contingents, or else to devise an adequate logic and semantics for which bivalence fails, and lastly the challenge posed by relativistic physics. Before we address these issues, however, we will first consider GBT's main competitors – presentism, and static and dynamic permanentism – and propose novel characterisations of these views that make their theoretical commitments more readily comparable with those undertaken by proponents of GBT.

Chapter 5
The Other Contenders

Abstract In this chapter we offer novel characterisations of presentism and permanentism which, or so we argue, significantly improve upon extant accounts. In particular, we show that, given the availability of these characterisations, neither presentism nor dynamic permanentism needs to invoke any substantial notion of presentness. In Sect. 5.1 we rehearse T. Williamson's misgivings about the use of the notion of presentness in attempts to articulate presentism. While Williamson takes these misgivings to be sufficient to discard presentism, in Sect. 5.2 we show that the view allows for its systematic reformulation solely in terms of tensed quantification, temporal operators and a predicate for times. In Sect. 5.3, after giving a characterisation of static permanentism and critically discussing R. Cameron's recent account of the Moving Spotlight Theory, we offer an equally lean formulation of dynamic permanentism solely in terms of temporal operators and a tensed proposition true at one time only.

In this chapter, we offer characterisations of the two main competitors of GBT – presentism and permanentism – characterisations that make use of the same conceptual and logical tools introduced in Chaps. 1, 2 and 3, or natural extensions thereof. In this way, we aim to show that GBT, as formulated in Chap. 4, is readily comparable with the other contenders, and pave the way to a systematic discussion of the relative merits and shortcomings of either of these different types of views.

We do not here undertake the task of discussing these relative merits and shortcomings, though. The purely descriptive task proves difficult enough. One reason for this is that, unlike GBT, presentism, as the main temporaryist alternative to GBT, is naturally expected to require the notion of *presentness* for its proper formulation and so to call for additional conceptual resources.

Permanentism, by contrast, would appear to be straightforwardly expressible using no more than temporal operators, quantification and identity:

(PER) Always, $\forall x$ Always, E!x

where, to recall, (D5) defined 'E!m' as '$\exists x(m = x)$'. However, permanentism comes in two basic varieties, static and dynamic; and the map of competing views would not be complete without including characterisations of either variety. In Chap. 1, we

F. Correia, S. Rosenkranz, *Nothing To Come*, Synthese Library 395,
https://doi.org/10.1007/978-3-319-78704-6_5

already identified two principles, one the negation of the other, that capture the contrast between static and dynamic views:

(STA) Always, $\forall p$ Always, ($\mathcal{T}p \rightarrow$ Always, $\mathcal{T}p$)
(DYN) Sometimes, $\exists p$ Sometimes, ($\mathcal{T}p$ & Sometimes, $\neg\, \mathcal{T}p$)

where, to recall, '$\mathcal{T}$' is short for 'It is true *simpliciter* that', which operator we assumed as a primitive. It is therefore natural to identify static permanentism with (STA) & (PER), and dynamic permanentism with (DYN) & (PER). In fact, we saw in Chap. 1 that (STA) implies (PER), given two natural assumptions about $\mathcal{T}$:

$$\mathcal{T}\varphi \rightarrow \varphi$$
$$\text{E!}m \rightarrow \mathcal{T}\text{E!}m$$

Now, it is true that proponents of dynamic permanentism endorse both (DYN) and (PER), and hence a view that is in opposition to static permanentism, presentism and GBT; and yet, if merely characterised in terms of (DYN) and (PER), dynamic permanentism would seem to remain in an important sense underdescribed. Dynamic theories of time typically commit to the idea that there is constant temporal change – that time constantly passes – where this is supposed to be rooted in an ongoing change in what facts there are, rather than being a mere projection. As such, dynamic theories are under the obligation to explain the nature of that ongoing factual change. However, (DYN) alone at best implies that there is some such change in what facts there are, without yet providing us with any insight into its nature. So more must be said in order to discharge the obligation. GBT, for instance, discharges this obligation when it identifies temporal passage with the constant variation in what has just become. Given their commitment to (PER), this is of course no option for dynamic permanentists.

The question accordingly is how proponents of dynamic permanentism can account for temporal passage. It is a natural enough thought that, to this end, dynamic permanentism will eventually have to invoke the notion of presentness or cognate notions. The classical *Moving Spotlight Theory*, to be examined below, does precisely that: according to this view, time passes as different times become present.

It is, of course, consistent with the framework developed in Chaps. 1, 2 and 3 that a view be characterised by further principles governing a distinguished set of predicates like, e.g., 'is present'. However, as Williamson (2013) has recently argued, the relevant notion of presentness remains too elusive to be at the service of metaphysical theorizing. Although these criticisms, to be reviewed in due course, are primarily directed against presentism, they carry over to the Moving Spotlight Theory, at least in its classical formulation.

As we shall see, to say that the notion of presentness is too elusive to be at the service of metaphysical theorizing is not to preclude that these theories can, at some stage, introduce such a notion. However, as we shall argue, to this end, the respective theories must already be in place; and there is then no saying that the notion of presentness they introduce is one and the same, so that one could ask which party gives a more faithful account of it.

As we argue in detail, and appearances notwithstanding, neither presentism nor dynamic permanentism is in need of any substantial notion of presentness. Thus, it transpires that presentism can instead be formulated using the same austere resources we already used to formulate GBT. Dynamic permanentism ultimately needs further conceptual tools – for example, as we will suggest later, metric temporal operators. Such metric operators are familiar enough, even if they made no appearance in Chaps. 1, 2 and 3. There is no principled reason to think that the framework developed there cannot systematically be expanded so as to include principles governing such operators.

The plan for this chapter is accordingly as follows. First – and here we closely follow Williamson (2013) – we identify a problem for the standard characterisation of presentism that accrues from its use of 'is present'. We then go on to give an alternative formulation of presentism that, by using only temporal operators, quantification and identity, proves congenial to the way in which we conceived GBT in Chap. 4. Subsequently, we show which of the temporal Barcan formulas, and converse Barcan formulas, respectively hold or fail to hold on this view.

In the second part of this chapter, we turn to the characterisation of dynamic permanentism. To begin with, we give a formulation of the classical Moving Spotlight Theory and show that while it escapes hostile takeover by eternalists, it nonetheless succumbs to the Williamsonian criticism that the relevant notion of presentness remains too elusive. We then reconstruct an alternative to the classical Moving Spotlight Theory that is inspired by Williamson himself, and argue that it too is fraught with difficulties. Next we turn to Ross Cameron's recent version of the Moving Spotlight Theory, critically review its essentials, and offer a diagnosis of why it ultimately remains unconvincing.

In the light of these criticisms, we then devise an alternative characterisation of dynamic permanentism that is immune to the kinds of problems that beset the versions examined thus far. It turns out that dynamic permanentism, as thus characterised, is entailed by all the versions of the Moving Spotlight Theory so far reviewed and hence can be seen as their defensible common core. More importantly in this context, the version of dynamic permanentism we end up with makes no use of the notion of presentness or cognate notions but can be expressed using only very minimal resources.

The upshot of the discussion then is that the pervasive thought, that dynamic theories of time must appeal, in some way or other and at some stage or other, to the notion of presentness or cognate notions, is mistaken across the board. The framework developed in Chaps. 1, 2 and 3 – or a suitable, but still very natural expansion of it – is thereby shown to be sufficient in order to allow for the systematic formulation and comparison of competing theories of time.

5.1 The Problem with Presentness[1]

We do not know who first started the rumour, but it is frequently said that presentism – somewhat carelessly cast as the thesis that everything is present – is either trivially true or clearly false, depending on how the quantifier 'everything' is being understood. Thus it is contended that if 'everything' here means the same as 'everything present', presentism is trivially true, whereas if 'everything' here means the same as 'everything past, present or future', presentism is clearly false (cf. Crisp 2004 for some pertinent references).

On the face of it, 'everything' simply means *everything* and does not mean anything else. However, it is a familiar phenomenon that uses of 'everything' are subject to implicit, contextually determined scope restrictions; and this may be so even in contexts in which one discusses ontology. There is no evident reason to believe, though, that presentists are unaware of the phenomenon. More specifically, there is no evident reason to believe that presentists are unaware of any implicit, contextually determined scope restriction that would trivialise their thesis. Yet, if they were aware of any such restriction in a given context, they would hardly, in that context, put forth their claim as a debatable thesis in need of defence.

Let us accordingly suppose, merely for the sake of argument, that in the particular context at hand, presentists use 'everything' to mean *everything past, present or future*. It does not follow that in asserting 'Everything is present', they thereby say something that is clearly false. To say that every black or non-black raven is black is a perfectly sound way of saying that the only ravens that exist are black – and this would remain to be so even if, throughout, we replaced 'black' and 'non-black' by 'partially black' and 'partially non-black' respectively. Similarly, to say that everything that is past, present or future is present is a perfectly sound way of saying that the only things in time that exist are present – and this will remain to be so even if it is assumed that what is present may also be past or future. Of course, once it is conceded that there are dinosaurs, that thesis must be considered falsified, just as 'Every black or non-black raven is black' will count as false if there are uniformly white ravens. Hence all the burden is on the claim that there are such things as dinosaurs; and every presentist in their right mind will deny this sort of claim. What they will be happy to accept, by contrast, is that in the past, there were dinosaurs. But they will hasten to add that this in no way implies that there are dinosaurs or that there are things that, in the past, were dinosaurs. (For more on this diagnosis, see Crisp 2004.)

The second horn of the alleged dilemma is sometimes put in terms of an *eternalist quantifier*, as if such a thing existed. It is then argued that, accordingly understood, 'Everything is present' is clearly false because the eternalist quantifier also ranges over dinosaurs. We may cast this line of thought as follows: 'Insofar as it is uncontroversial that in the past there were dinosaurs, the eternalist quantifier will range over dinosaurs. And insofar as it is uncontroversial that no dinosaurs are pres-

[1]This section and the next are based on Correia and Rosenkranz (2015b), but contain new material.

ent, “Everything is present” will accordingly come out false once “everything” is taken to express that eternalist quantifier!’

However, if there are no dinosaurs because everything is present and no dinosaurs are, then not even an ‘eternalist’ quantifier manages to range over dinosaurs. We hear it being replied that any way of saying something true by saying that there are no dinosaurs for these kinds of reasons must deploy a *presentist quantifier* (as if such a thing existed), and that this takes us right back to the first horn. Once ‘everything’, as used by the presentist, is understood to mean the same as ‘everything present’, and ‘nothing’, as thus used, is correspondingly understood to mean the same as ‘nothing present’, then clearly and unspectacularly, nothing is a dinosaur in this sense of ‘nothing’.

This reply is as good as saying that it is trivial that no raven is uniformly white if by ‘no raven’ we mean the same as ‘no black raven’. Of course, this is *not* what we mean by ‘no raven’ when we assert ‘No raven is uniformly white’. Why on earth should the presentist feel any corresponding pressure to concede that what she really means to be saying when asserting ‘Everything is present’ is that everything present is present? We know of no remotely plausible argument to this effect.[2] If her assertion is nonetheless deemed false once her use of ‘everything’ is taken to mean something less restrictive, then this requires argument. It anyway will not do to simply insist that if ‘everything’ does not mean the same as ‘everything present’, there will be dinosaurs for it to range over.

One mistaken assumption here is that there are different quantifiers to choose from in order to interpret the claim ‘Everything is present’. There is only one candidate quantifier to properly interpret the presentist’s use of ‘everything’, and that is the quantifier we all express whenever we use ‘everything’, with or without explicit or contextually determined scope restriction. Presentists, permanentists and proponents of GBT alike use the very same quantifier with exactly the same intended meaning and in the very same context, and yet make conflicting claims about its range. This is by far more fruitful a rendition of the debate that these parties engage in than any interpretation that posits unwanted and unwarranted ambiguities or contextual variations that make its participants enunciate trivialities or blatant falsehoods. Far from being charitable, any such interpretation makes those engaged in the debate look like fools; and one cannot stipulate things into existence by one’s conceptual choices either.[3]

Indeed, that quantification is univocal has been our assumption since Chap. 2. There we laid down principles governing the quantifiers that yield a minimal quantification theory meant to be acceptable to all parties involved. As was noted,

[2] One source of confusion might be the mistaken thought that insofar as quantification is tensed, as we have assumed it is, ‘E!*m*’ proves equivalent to ‘*m* timelessly exists & *m* is present’ (see, for instance, Meyer 2013b). This would be to misunderstand the significance of the present tense. For instance, when one says ‘The crisis is over’, one does not thereby affirm that the crisis is timelessly over and also present; and there is no reason to think that the tensedness of ‘E!*m*’ differs so radically from the tensedness of predication.

[3] Cf. Putnam (1962) on the abuse of the analytic-synthetic distinction.

once permanentism's key thesis is added to the mix, that theory will expand to classical quantification theory. But we are well-advised not to think that, by adding it, one changes the meaning of those other principles. Similarly, by laying down GBT's basic tenets, we were best not seen as effecting any meaning change: principles such as the principle that always there is something new, or the principle that nothing will ever be lost, express substantial metaphysical claims that would not become any less substantial if, quite inappropriately, they were declared meaning postulates. The same should hold for whatever metaphysical claim about existence presentists can ultimately be said to endorse.

Now in fact, presentists are unlikely to commit to the unqualified claim that everything is present. For, what about abstract objects like numbers? Are presentists *qua* presentists committed to denying that there are any abstract objects? Hardly. But then, if they were to say that everything – the number 453 included – is present, what conception of being present would they presuppose? If abstract objects exist, they presently exist, since in general, if φ then presently φ. By contrast, abstract objects are not present in any more demanding sense in which being present implies being about or being amongst us. So it is tempting to construe any such wholesale claim in terms of present existence rather than anything more demanding. On this reading, to say that everything is present is just to say that everything presently exists; and here 'everything' can be understood to include abstract objects within its range (cf. Williamson 2013: 24).

However, this now is a claim that eternalists are likewise happy to endorse. If dinosaurs exist, as eternalists claim, then they presently exist. Accordingly, presentists had better not claim of everything, abstract or non-abstract alike, that it is present, because – as we have just seen – this would force a reading of 'is present' that makes that claim trivially true. Instead, they should be understood to be saying no more than that everything *in time* is present, where this now involves a genuine and explicit restriction rather than a different sort of quantifier. This still leaves the question of what is here meant by 'is present'.

In reply, one might insist that 'present' simply means *present* and nothing else, and so in particular does not mean *presently existing*. However, the view according to which always everything always exists and things that once were dinosaurs are still about – albeit in a non-concrete way inconsistent with their presently being dinosaurs – shows that presentists cannot rest content with such deflationary a characterisation (cf. Williamson 2013: 7–8). For all that has been said, a proponent of such a dynamic permanentist view might likewise endorse that everything in time is present while conceding that absolutely nothing is a dinosaur. And yet, she will also insist that there are non-concrete things that once were dinosaurs – a claim that presentists would want to reject.

At the same time, however, presentists cannot successfully rephrase their claim in terms of being concrete either. For, even according to some eternalists, everything in time may be said to be concrete, including dinosaurs located in the past; and since if φ then presently φ, for such eternalists, everything in time will also presently be concrete, even if some such things are only located at remote times. Thus, if there is any dilemma at all that presentists face, it has to do with the intended reading of 'is present' rather than the intended reading of 'everything'.

The presentist may of course try out combinations of these claims, e.g. by saying that everything in time is both concrete and presently about – that it is, to borrow Moore's phrase, presently 'to be met with in space' (Moore 1939). Ultimately, however, it is far from clear why her presentism should commit the presentist to an ontology of things in time exclusively composed of concrete, or presently spatially located, things (and, perhaps, sets and fusions thereof). After all, if it at all makes sense to speak of things that are neither abstract nor concrete, why shouldn't some of the present things be such – even if ex-dinosaurs are not among them? And do all one's mental states, for instance, have to be spatially located in order to exist (Williamson 2013: 24)?

Yet another attempt to give sense to the relevant notion of presentness, suggested by Cian Dorr and Jeremy Goodman, proceeds from the observation that φ iff presently φ, and contends that to be present, in the sought-after sense, is to be located at the one-and-only *accurate* time (Dorr and Goodman forthcoming). Here, a time m is accurate iff for all φ, φ iff At m, φ. Given that there is a unique accurate time, the present time would then be the accurate time. The problem here is that, unless the propositional quantifier ranges over propositions that already contain attributions of presentness, which would make the characterisation intolerably circular, there is no evident conceptual reason to rule out that more than one time is accurate in the sense laid down.

Williamson may be right, then, that we need a fresh start in order to articulate the debate among different views on time and existence (Williamson 2013: 25). However, his conclusion that we should articulate the debate in such a way that the traditional oppositions no longer figure, seems to us to be both highly implausible and premature. The availability of the characterisation of GBT that we devised in Chap. 4 already gives proof of this. As we shall argue in the next section, presentism can likewise perspicuously be formulated without any appeal to the notion of being present – or the notion of being concrete, for that matter.

5.2 Presentism without Presentness

Let us say, as before, that *m is new* if, and only if, m exists and always in the past, m does not exist. Correspondingly, let us say that *m is obsolescent* if, and only if, m exists and always in the future, m does not exist. Given these definitions, let us stipulatively define *being one-off* as follows:

$$m \text{ is one-off} \equiv_{\text{df}} m \text{ is new } \& \ m \text{ is obsolescent}$$

In a first go, one might be tempted to characterise presentism as follows:

(Pres**) x is in time $\rightarrow$ x is one-off

where, to recall,

(D6) m is in time $\equiv_{df}$ Sometimes, $\exists x(m \mathrm{L} x)$

Accordingly, it follows from (Pres**) that always, every time is one-off, which is something that presentists would indeed want to say. However, (Pres**) has the unpalatable consequence that every resident of time is instantaneous, where, to recall,

(D8) m is instantaneous $\equiv_{df}$ m is in time & Always, $\forall x$ Always, $\forall y(m \mathrm{L} x \;\&\; m \mathrm{L} y \rightarrow x = y)$

and *being a resident of time* is defined thus:

(D7) $\mathrm{R}m \equiv_{df} m$ is in time & Always, $\forall x$ Always, $(m \mathrm{L} x \rightarrow (\mathrm{E}!x \rightarrow \mathrm{E}!m))$ & Always, $(\mathrm{E}!m \rightarrow \exists y(m \mathrm{L} y))$

Proof: Suppose Rm and m is not instantaneous, i.e. Sometimes, $\exists x$ Sometimes, $\exists y(m \mathrm{L} x \;\&\; m \mathrm{L} y \;\&\; x \neq y)$. By (A25) and (A32), then, Sometimes, $\exists x$ Sometimes, $\exists y(m \mathrm{L} x \;\&\; m \mathrm{L} y \;\&\; (x \prec y \vee y \prec x))$. By (T10) and (T11), the following are theorems:

(T17) $x \prec y \rightarrow ((\mathrm{At}\; y, \varphi) \rightarrow \mathrm{At}\; x, \mathrm{F}\varphi)$ (with y not free in φ)
(T18) $y \prec x \rightarrow ((\mathrm{At}\; y, \varphi) \rightarrow \mathrm{At}\; x, \mathrm{P}\varphi)$ (with y not free in φ)

Now recall

(A23) $\mathrm{T}x \rightarrow \mathrm{At}\; x, \mathrm{E}!x$

By (A23), (T17), (T18) and (D8), it follows that Sometimes, $\exists x$ Sometimes, At x, $(\mathrm{E}!m \;\&\; (\mathrm{FE}!m \vee \mathrm{PE}!m))$. Hence using (T2), one can show that Sometimes, $(\mathrm{E}!m \;\&\; (\mathrm{FE}!m \vee \mathrm{PE}!m))$, which latter conflicts with the assumption that m is one-off. □

Most presentists, however, would want to retain belief in residents of time that are located at distinct times. This suggests that presentists should instead endorse the following, weaker claim:

(Pres*) $\exists x(\mathrm{T}x \;\&\; x$ is one-off$)$

Unlike (Pres**), (Pres*) leaves room for non-instantaneous residents of time. However, (Pres*) alone will not do. By (D7), there cannot exist residents of time that will only ever be located at some time in the future, or have only ever been located at some time in the past, without there presently being a time at which they are located. However, this much still is compatible with there being residents of time – e.g. dinosaurs – such that the only times at which these things are located are times which exist and already existed in the past. This is something that presentists will want to rule out. That is to say, although (Pres**) was too strong a claim, presentists will still want to ensure something that (Pres**) entails, *viz.* that always, all times are one-off. Yet, (Pres*) only affirms the existence of a one-off time, which is evidently compatible with the existence of other times that are not one-off but already existed in the past, and so compatible with the existence of dinosaurs.

Come to think of it, the mere idea that there presently exist *distinct* times – and so by the totality axiom (A32), times that stand in relations of precedence – offends against standard presentist ideology – at least if, as we have presumed from the onset, quantification over times is more than quantification over purely abstract things. It is therefore anyway incumbent upon presentists to endorse

(Pres) $(\mathrm{E}!x \,\&\, \mathrm{T}x \,\&\, \mathrm{E}!y \,\&\, \mathrm{T}y) \rightarrow x = y$

(Pres) entails (Pres*). To see this, we first prove that the following two principles can be derived from (Pres):

(P2) $\mathrm{T}x \rightarrow \mathrm{At}\ x, \mathrm{H}\neg\mathrm{E}!x$

(P3) $\mathrm{T}x \rightarrow \mathrm{At}\ x, \mathrm{G}\neg\mathrm{E}!x$

Proof: Assume At x, PE!x. By (A33), then, Sometimes, $\exists y(y < x \,\&\, \mathrm{At}\ y, \mathrm{E}!x)$. By (A23), (A28) and the irreflexivity of precedence, it follows that Sometimes, $\exists y(y \neq x \,\&\, \mathrm{At}\ y, (\mathrm{E}!x \,\&\, \mathrm{T}x \,\&\, \mathrm{E}!y \,\&\, \mathrm{T}y))$. But then, by (Pres), Sometimes, $\exists y(y \neq x \,\&\, \mathrm{At}\ y, x = y)$, which is impossible. Hence, (Pres) entails (P2). One can show in a similar way that (Pres) entails (P3), by invoking (A34) instead of (A33). □

Accordingly, (Pres) entails one of the two axioms characterising GBT. Next we show that (Pres*) follows from (P2) and (P3).

Proof: Suppose for *reductio* that $\forall x(\mathrm{T}x \rightarrow (\mathrm{PE}!x \vee \mathrm{FE}!x))$. By (T6), it follows that $\exists x(\mathrm{T}x \,\&\, (\mathrm{PE}!x \vee \mathrm{FE}!x))$. By (A22) and (A23), we then get: $\exists x \exists y\ \mathrm{At}\ y, (\mathrm{T}x \,\&\, \mathrm{E}!x \,\&\, \mathrm{T}y \,\&\, \mathrm{E}!y \,\&\, (\mathrm{PE}!x \vee \mathrm{FE}!x))$, and so by (Pres): $\exists x\ \mathrm{At}\ x, (\mathrm{PE}!x \vee \mathrm{FE}!x)$. But this is excluded by (P3) and (P2). □

(Pres) accordingly entails, not only that always all times are one-off, but also that every resident of time is located at the only time that there is, where this time is presently one-off. Accordingly, there exist no residents of time such that any time at which they are located already existed in the past. If there were dinosaurs, they would be things of that kind. Hence, according to (Pres), there are no dinosaurs. Similarly, Martian outposts, if such there be, would be things such that any time at which they are located would exist in the future. Similarly, then, according to (Pres), there are no Martian outposts. Entailing (P3), (Pres) conflicts with GBT, and entailing (P2), it likewise conflicts with permanentism. (Pres) is therefore a good candidate for capturing the presentists' view. It turns out that, conversely, (P2) and (P3) jointly entail (Pres).

Proof: Let us first establish that (P2) and (P3) jointly entail (Tx & At z, E!x) → $x = z$. Suppose for *reductio* that (Tx & At z, E!x) & $x \neq z$. Then by (A32), (At z, E!x) & ($x \prec z \vee z \prec x$). By (T17), (T18), (P2) and (P3), this is impossible. Let us now establish that (P2) and (P3) entail (Pres). Assume (P2) and (P3), and suppose for *reductio* that (E!x & Tx & E!y & Ty) & $x \neq y$. By (T5) we then have: ∃z At z, (E!x & E!y). By the result established above, then, $x = z$ & $y = z$, and so $x = y$. Contradiction. □

Consequently, there are two equivalent candidate characterisations of presentism. In fact, characterising presentism by the combination of (P2) and (P3) has the distinctive advantage of making presentism more easily comparable with GBT and, as we shall see, also with permanentism.

On the basis of (Pres), we can show that the one and only time it posits is accurate. To begin with, recall

(A22) $\forall x(\mathrm{T}x \rightarrow \mathrm{At}\ x, \varphi) \rightarrow \varphi$ (with x not free in φ)
(T5) $\varphi \rightarrow \exists x\ \mathrm{At}\ x, \varphi$ (with x not free in φ)

From this and (Pres), we can straightforwardly derive.

$$\forall x(\mathrm{T}x \rightarrow \mathrm{At}\ x, \varphi) \leftrightarrow \varphi \quad \text{(with } x \text{ not free in } \varphi)$$

Consequently, the one and only time will *a fortiori* be accurate. Presentists can now likewise offer the following partial definition of being present:

(D13) Rm → (m is present ≡ ∃x(x is one-off & m L x))

How this partial definition might be used to arrive at a full definition of being present will *inter alia* depend on the existence conditions of other types of things in time that one's overall theory might countenance. But this question need not detain us here. What is important to note is that it would be a mistake to see the availability of (D13) as a rehabilitation of the notion of presentness as a serviceable tool for metaphysical theorizing. As the discussion in the previous section made clear, we do not have a previously well understood notion of being present that we might appeal to in this context; and (D13) itself cannot be said to elucidate the notion of being present because it cannot be viewed as a partial or total *analysis* of it. If it were such an analysis, GBT and any version of dynamic permanentism framed in terms of presentness would be *analytically* false. At best, (D13) is a partial *real* definition of being present, which stands and falls with the theory that issues it. If there is no one-off time that is the only time that there is, so that presentism proves false, then no resident of time satisfies 'is present' as this term is characterised by (D13); and plausibly, if no resident of time does, nothing in time does. In this sense, the introduction of 'is present' by means of (D13) is a mere after-thought. The same applies to the following, alternative partial definition of being present that presentists might offer:

(D14) $\mathrm{R}m \rightarrow (m \text{ is present} \equiv \exists x(m \mathrm{\ L\ } x))$

This diagnosis does not change substantially, once we reflect that for presentists, a time is new iff it is one-off, and that they may accordingly agree with proponents of GBT on the following partial definition instead:

(D15) $\mathrm{R}m \rightarrow (m \text{ is present} \equiv \exists x(x \text{ is new } \& \ m \mathrm{\ L\ } x))$

This partial definition still is available only on presentism or GBT, and failing our command of any previously well-understood notion of presentness, cannot be taken to vindicate the assumption that there is such a notion which we might apply independently from endorsing one or the other of these theories.

Insofar as there are any times at all, as is entailed by (A22), very plausibly, the indexical 'now' never fails to refer. Given these assumptions, it is trivial that, presently, 'now' refers to now. According to presentism, always, 'now' refers, if to anything, to the one-off time that is the only time that there is. It follows that now is the one-off time that is the only time that there is; and similarly, it follows that now is the time at which every resident of time is located. This should allay even the last doubts about whether presentism, as here characterised, is faithful to standard presentist ideology, despite the fact that its current formulation nowhere relies on the notion of presentness or cognate notions, but instead only deploys temporal operators, quantification, identity and the notion of being a time.

We submit that (Pres), or equivalently, the combination of (P2) and (P3), is all that presentists need. By endorsing these principles, presentists can skirt any problems – conceptual, theoretical, or dialectical – that accrue from the use of 'is present', or cognates, in attempts to formulate their view. In particular, they need no longer concern themselves with the task of providing an interpretation of that term that both ensures that their core claims are neither trivially true nor obviously false and forestalls any hostile takeover by their opponents.

On this rendition of the view, presentism entails that always there is a time that is the only time there is, and that never existed before and will never exist thereafter. Presentism thus remains in opposition to permanentism and GBT, as on neither of these views is there ever anything that is obsolescent. Presentism likewise entails that always, every resident of time is located at the unique time that presently exists, thereby ensuring that presently there are no dinosaurs and no Martian outposts. In accordance with what most presentists would want to say, it nonetheless leaves room for the existence of non-instantaneous residents of time that in the past were located at the unique time that then existed or that, in the future, will be located at the unique time that will then exist.[4]

[4] Presentism as characterised allows for temporally extended things *with continuous existences*, e.g. football matches, and can validate what we ordinarily say about them, making use of metric tense-operators. 'The match will last for 90 minutes' can be paraphrased as '90 minutes hence, (the match takes place & $\forall n(n \leq 90 \leftrightarrow n$ minutes ago, the match takes place))'. Note that here we quantify over numbers and not times.

Presentism as stated allows for non-concrete residents of time, as long as they are located at the only time that presently exists. But presentists are neither committed to the existence of any non-concrete things in time nor do they have any need for such things in order to make their ontological thesis cohere with the observation that, in the past, there were dinosaurs. True: while presentism as stated rules out dinosaurs – since no dinosaur is located now – it does not likewise rule out ex-dinosaurs, i.e. things that are not dinosaurs but once were. But to the extent that there is no reason at all to believe in ex-dinosaurs – let alone, in ex-dinosaurs *located now* – as long as there is no independent reason to believe the thesis that always everything always in the future exists – a thesis that presentists are anyway committed to reject – this problem, insofar as it is a problem at all, proves purely academic.

We close this section by examining which of the Barcan formulas and their converses presentists are bound to accept, and which of them they are bound to reject. Since presentism is a version of temporaryism which is committed to (P2), on this view, the Barcan formula for 'F', the converse Barcan formulas for 'P', 'H' and 'Sometimes' and operators of the form 'At m', as well as $(\mathrm{CBF_{At}})^{\forall}$ all fail for the very same reasons already mentioned in Chap. 4. As temporaryists, presentists must likewise reject the Barcan formula for 'Sometimes' and the converse Barcan formula for 'Always' – again for the reasons outlined in Chap. 4. The argument we devised there for the failure of the Barcan formulas for 'H', 'G' and 'Always' did not depend on any specific metaphysics of time; accordingly, these formulas likewise fail on the presentists' view.

What remains to be shown is whether presentism is consistent with the Barcan formulas for 'P' and operators of the form 'At m', and with the converse Barcan formulas for 'F' and 'G'. Let us consider the Barcan formula for 'P' first, i.e.

$(\mathrm{BF_P})$ $\quad \mathrm{P}\exists x\varphi \rightarrow \exists x\mathrm{P}\varphi$

Let φ be 'x is obsolescent'. Provided that time has not just begun, according to presentism, $\mathrm{P}\exists x\varphi$ holds. However, since $\mathrm{PG}\psi$ entails ψ, $\exists x\mathrm{P}\varphi$ fails. Consequently, given presentism, $(\mathrm{BF_P})$ cannot be assumed. $(\mathrm{BF_{At}})$ fails on presentism for the very same reason it failed on GBT (see Chap. 4). By contrast,

$(\mathrm{BF_{At}})^{\forall}$ $\quad \forall x((\mathrm{At}\ x, \exists y\varphi) \rightarrow \exists y\ \mathrm{At}\ x, \varphi)$ (with x and y distinct variables)

can be proved for presentism as follows:

Proof: Using (T2) and (T12), we derive '$\mathrm{E!}x \rightarrow ((\mathrm{At}\ x, \exists y\varphi) \rightarrow$ Sometimes, (x is new & $\exists y\varphi$))'. Given (A2) and the definition of being new, '$\mathrm{E!}x \rightarrow \neg\mathrm{F}$($x$ is new)' is a theorem. Since '$\mathrm{T}x$ & $\mathrm{E!}x \rightarrow x$ is new' is a theorem of presentism (this follows from (T5), (A23), (Pres) and (P2)), we can then derive '$\mathrm{E!}x \rightarrow ((\mathrm{At}\ x, \exists y\varphi) \rightarrow$ (x is new & $\exists y\varphi$))'. Now since x and y are distinct variables, '(x is new & $\exists y\varphi$)' entails '$\exists y$(x is new & φ)'. Since the latter yields '$\exists y$ Sometimes, (x is new & φ)', we can derive '$\mathrm{E!}x \rightarrow ((\mathrm{At}\ x, \exists y\varphi) \rightarrow \exists y$ Sometimes, (x is new & φ))'. Using (T16) we can then derive '$\mathrm{E!}x \rightarrow ((\mathrm{At}\ x, \exists y\varphi) \rightarrow \exists y\ \mathrm{At}\ x, \varphi)$', which entails $(\mathrm{BF_{At}})^{\forall}$. □

Table 5.1 Barcan and converse Barcan: presentism and GBT

	Barcan Formula		**Converse Barcan Formula**	
	Presentism	GBT	Presentism	GBT
P	No	Yes	No	No
B	No	No	No	No
F	No	No	No	Yes
G	No	No	No	Yes
Sometimes	No	No	No	No
Always	No	No	No	No
At m	No	No	No	No
	$\mathbf{(BF_{At})^{\forall}}$		$\mathbf{(CBF_{At})^{\forall}}$	
	Presentism	GBT	Presentism	GBT
	Yes	Yes	No	No

We next turn to the converse Barcan formula for 'G', i.e.

$$(\mathrm{CBF_G}) \qquad \exists x \mathrm{G}\varphi \rightarrow \mathrm{G}\exists x\varphi$$

Let φ be 'x does not exist'. According to presentism, $\exists x\mathrm{G}\varphi$ holds. Given that time has not come to an end, $\mathrm{G}\exists x\varphi$ fails. Under the same replacement of φ, given presentism,

$$(\mathrm{CBF_F}) \qquad \exists x \mathrm{F}\varphi \rightarrow \mathrm{F}\exists x\varphi$$

fails. (Note that if time has come to an end, ($\mathrm{CBF_F}$) is true already because its antecedent is false.) Hence, on the presentists' view, neither ($\mathrm{CBF_G}$) nor ($\mathrm{CBF_F}$) can be assumed to hold.

Table 5.1 summarises these commitments of presentism and juxtaposes them with those incurred by proponents of GBT, thereby expanding Table 4.1 from Chap. 4.

5.3 Dynamic Permanentism

Permanentism is the view according to which always everything always exists. We have so far expressed this by

$$(\mathrm{PER}) \qquad \text{Always}, \forall x \text{ Always}, \mathrm{E}!x$$

Keeping in mind that all of our principles must be understood in such a way that they allow prefixing by any combination of the universal quantifier and 'Always', a more elegant way of expressing this same thought is

(PER′) $E!x$

Each of (PER) and (PER′) is equivalent to the conjunction of the following two principles:

(P1) $E!x \rightarrow GE!x$
(P4) $E!x \rightarrow HE!x$

This equivalent characterisation has the distinctive advantage of making permanentism more readily comparable with GBT and presentism as these views have here been formulated. Thus, (P1) is incompatible with presentism which latter entails

$$Tx \,\&\, E!x \rightarrow G\neg E!x$$

From this, by (P1), we would get: $Tx \,\&\, E!x \rightarrow G(E!x \,\&\, \neg E!x)$, and hence: $Tx \,\&\, E!x \rightarrow G\bot$, where $\bot$ is a contradiction in which x does not occur free. From (T6), we anyway have $\exists xTx$, so that we can now derive $G\bot$, and hence: Always, $G\bot$. But this is incompatible with (A14).

Similarly, (P4) is incompatible with GBT which latter entails

$$Tx \,\&\, E!x \rightarrow H\neg E!x$$

From this, by (P4), we would get: $Tx \,\&\, E!x \rightarrow H(E!x \,\&\, \neg E!x)$, and so: $Tx \,\&\, E!x \rightarrow H\bot$, where $\bot$ is a contradiction in which x does not occur free. From (T6) we anyway have $\exists xTx$, so that we can now derive $H\bot$, and hence: Always, $H\bot$. But this is incompatible with (A14).

Permanentism comes in two basic varieties: static permanentism and dynamic permanentism. Static permanentism is commonly called 'eternalism'. The label is apt, since there are, on the static view, no tensed truths *simpliciter*: all truths *simpliciter* – and not just those concerning existence – are eternal truths. Dynamic permanentists agree that truths about existence are eternal. But they nonetheless insist on the dynamic nature of temporal reality and find it manifest in the variation of what can truly be predicated of permanently existing things.

Throughout we have assumed that truths about existence, in the sense of (D5), are truths *simpliciter*. Evidently, the same cannot be assumed for all other truths; and since even proponents of static permanentism need not, *eo ipso*, have any qualms about the truth-aptitude of tensed sentences, formulation of the contrast between static and dynamic permanentism ultimately requires appeal to the notion of truth *simpliciter*.

Dynamic permanentists thus endorse both (PER) and (DYN). But for the reasons mentioned in the introduction to the present chapter, this minimal characterisation, though perhaps good enough for purely classificatory purposes, would not seem to do sufficient justice to the theoretical obligations that philosophers of time who deny (STA) incur. One of these theoretical obligations certainly is to explain in what constant temporal change consists – what facts continually change as time goes by. Temporaryists, like presentists and proponents of GBT, discharge this obligation by postulating a constant change in what exists. Dynamic permanentists will have to do so in other ways.

(PER) not only yields classical quantification theory. It also has the consequence that dynamic permanentists cannot exploit Broad's insight that temporal relations like precedence, though rigid, may not themselves be eternal – *viz.* when these relations are construed as existence-entailing, while what exists can be said to change with time. If, on the other hand, these relations are *not* construed as existence-entailing, then it would seem that their rigidity is quite sufficient for their being eternal; and then the only way to allow for variation in truths about what precedes what would again be to adopt a temporaryist ontology and claim that what exists changes with time. However, no one would want to deny that precedence is rigid. So either way, given (PER), truths about what precedes what will be eternal. Similarly, no one would want to deny that location is rigid; and so given (PER), truths about what is located at what time will likewise be eternal. In other words, then, dynamic permanentists cannot hope to express constant temporal change in terms of McTaggart's B-relations. It is therefore natural for them to turn to McTaggart's A-properties instead, when formulating their preferred principle of temporal change.

This is what proponents of the classical Moving Spotlight Theory do. According to this view, different times become present as time goes by. We might express this view as follows:

($M1_p$) $\exists x(\mathrm{T}x \,\&\, x \text{ is present} \,\&\, \mathrm{H}\neg(x \text{ is present}) \,\&\, \mathrm{G}\neg(x \text{ is present}) \,\&$
$\forall y(\mathrm{R}y \,\&\, y \text{ is present} \rightarrow y\ \mathrm{L}\ x))$

where, to recall, being a *resident of time* was defined as follows:

(D7) $\mathrm{R}m \equiv_{\mathrm{df}} m \text{ is in time} \,\&\, \text{Always, } \forall x \text{ Always, } (m\ \mathrm{L}\ x \rightarrow (\mathrm{E!}x \rightarrow \mathrm{E!}m)) \,\&$
$\text{Always, } (\mathrm{E!}m \rightarrow \exists y(m\ \mathrm{L}\ y))$

It follows that always there is at most one time that is present. It likewise follows that a given time is only ever present once, and hence that always a distinct time is present. As such, ($M1_p$) is sufficient to rule out the untoward hypothesis, envisaged by Cameron (2015: 2), that the spotlight of presentness is, as it were, stuck or frozen. What ($M1_p$) does not yet guarantee, however, is that every time sometimes is present; so it is natural to add

($M2_p$) $\mathrm{T}x \rightarrow \text{Sometimes, } x \text{ is present}$

$(M1_p)$ and $(M2_p)$ are, of course, not all that proponents of the classical Moving Spotlight Theory characteristically claim to hold. To begin with, recall that presentists and proponents of GBT can ultimately agree on the following partial real definition of 'is present':

(D15) $Rm \rightarrow (m \text{ is present} \equiv \exists x(x \text{ is new } \& \ m \, L \, x))$

Given that, on both these views, what time is new changes with time, they can accordingly appropriate the letter of $(M1_p)$ and $(M2_p)$. However, we must not forget that $(M1_p)$ and $(M2_p)$ are supposed to be conjoined with (PER); and it then becomes abundantly clear that 'is present' cannot here be understood in any way that would cohere with the sense that (D15) confers on this term.

Accordingly, (PER) must be added to the mix. However, as Fine (2005) notes, $(M1_p)$ and $(M2_p)$ are truths that even eternalists would want to accept – the reason being that their commitment to an eternalist metaphysics does nothing to prevent them from engaging in ordinary tensed talk. As far as ordinary tensed talk goes, $(M1_p)$ and $(M2_p)$ indeed sound truistic. Consequently, even the conjunction of $(M1_p)$ and $(M2_p)$ with (PER) does not yet suffice for dynamic permanentism.

It is for this reason that proponents of the Moving Spotlight Theory had better add the following commentary on their use of 'is present' that brings out their commitment to (DYN):

$$m \text{ is present} \rightarrow \mathscr{T}(m \text{ is present})$$

Once this commentary, on how $(M1_p)$ and $(M2_p)$ ought to be understood, is in place, the eternalist's hostile takeover can successfully be averted. However, with this commentary added, the view is now even more obviously subject to the kind of criticism we rehearsed in Sect. 5.1. $(M1_p)$ and $(M2_p)$ could, despite all Williamsonian worries about the notion of presentness, be conceived of as true even from the eternalist's perspective, to the extent that they in principle allowed for systematic reinterpretation in terms of a time-relative notion of presentness, *being present at*, which latter ultimately reduces to relations of location at times. Thus, from the eternalist's perspective, to say that always, there is a present time that has never been present before and will never be present again, is to lay claim to no fact other than that every time is present at itself and at no earlier or later time. But now that it is being assumed that if m is present, then $\mathscr{T}(m \text{ is present})$, this reinterpretation is of course foreclosed – just as intended – since we are now obliged to conceive of presentness as a monadic property that times can possess absolutely, i.e. without any further relativisation to temporal parameters, and that accordingly cannot be understood in terms of relations of *being present at*. But now the question rearises what this property is, which brings us right back to the quagmire that we outlined at the beginning of this chapter.

In the light of such problems, Williamson instead suggests conceiving of reality's dynamic nature, if any, as grounded in temporal shifts from non-concreteness to concreteness and in corresponding temporal shifts from concreteness to non-concreteness (Williamson 2013: 6–18, 24–25, 28–29). Thus, for example, at some moment in the closed interval running from my conception to my birth, I turned from being a non-concrete individual into a concrete one, and once I die, I will fall back into the realm of the non-concrete where I will stay forever after. Williamson nowhere provides a perspicuous account of concreteness. Instead, he writes in a footnote:

> The term 'concrete' is used informally throughout this book. For present purposes, we need not decide between various ways of making it precise (being material, being in space, being in time, having causes, having effects, …). (Williamson 2013: 6*n*)

In the light of his dismissal of theories that invoke presentness, this might strike one as quite a cavalier attitude to take, at least given how persistently the notion of concreteness makes its appearance in Williamson's book.

In any case, however, to the extent that dynamic permanentists must, just like their temporaryist opponents, account for constant variation across time, they are poorly advised to bank their account on births and deaths. Indeed, as Shoemaker (1969) has shown, there is nothing incoherent in the thought of time without physical change; and so even in a world without constant physical change, time might still constantly pass. Given that Williamson is quite happy to apply the notion of concreteness to times, it is therefore natural to model the type of view he suggests on ($M1_p$), systematically replacing 'is present' by 'is concrete':

($M1_c$) $\exists x(\mathrm{T}x\ \&\ x \text{ is concrete}\ \&\ \mathrm{H}\neg(x \text{ is concrete})\ \&\ \mathrm{G}\neg(x \text{ is concrete})\ \&$
$\forall y(\mathrm{R}y\ \&\ y \text{ is concrete} \rightarrow y\ \mathrm{L}\ x))$

In the same spirit as before, we will assume that ($M1_c$) is supplemented by

($M2_c$) $\mathrm{T}x \rightarrow$ Sometimes, x is concrete

and accompanied by the commentary that

$$m \text{ is concrete} \rightarrow \mathscr{T}(m \text{ is concrete})$$

However, while it is plausible to take the presentness of things in time to be explicable in terms of the presentness of some time at which they are located, when it comes to concreteness, the explanatory direction would anyway seem reversed. What it is for a time to be concrete cannot be rendered intelligible without recourse to the concreteness of the things located at it. This has, no doubt, something to do with the fact that we typically understand concrete individuals to be things 'to be met with in space', while it is far from clear what it would mean for a time to be spatially located. One may of course try to somehow reduce times in terms of the concrete things located at it, so that, in a sense, times are after all spatially located. But this does nothing to change the current diagnosis.

The problem now is that while it may be deemed a necessary condition for a time m to be concrete that all the things located at it are concrete (ignoring controversial types of mental phenomena), as long as none of those things is guaranteed to be instantaneous, all those things may both be concrete and located at another time *n* distinct from *m*. Given that ($M1_c$) entails that at most one time is ever concrete, the condition is therefore insufficient to explain what it is for a time to be concrete.

How, then, can it be guaranteed that always, for every time, amongst the things located at that time, some are instantaneous? It might be thought that the most straightforward way to issue such a guarantee is to identify the instantaneous things in question with snapshot distributions of matter across the spatial extension of the universe. By definition, instantaneous things in time are located at one time only:

(D8) m is instantaneous $\equiv_{df}$ m is in time & Always, $\forall x$ Always, $\forall y(m \text{ L } x \ \& \ m \text{ L } y \rightarrow x = y)$

However, on a permanentist view at least, it is far from clear why the same snapshot distribution of matter across the spatial extension of the universe could not recur at different times; and it is equally unclear why *this* possibility should be tied in any way to the possibility of time being *cyclical*.

The problem is akin to the one that already beset the idea of identifying the presentness of things in time as their location at the sole accurate time: just as it was unclear how to ensure that there are no distinct times at which all the same propositions hold – without, on pain of circularity, including propositions attributing presentness – it is unclear how to ensure that there are no distinct times at which the same concrete things are located – without, on pain of circularity, including uniquely concrete times. Accordingly, although it avoids any appeal to presentness, the view characterised by ($M1_c$) and ($M2_c$) is likewise fraught with difficulties. Time to move on.

In a recent monograph, Ross Cameron has argued for a dynamic permanentist view according to which the way concrete substances are, were, or will be, is fully determined by a combination of two kinds of factors: their present *age* and their *temporal distributional properties* (Cameron 2015: section 4.3). The temporal distributional property of a given concrete substance is understood to map out the career of that substance across its different ages and so to determine, for any age of that substance, how it is, was, and will be, at that age. While a concrete substance is said to always have the same temporal distributional property, plausibly, its age is supposed to vary with time, so that it always has a different age. Present-tensed truths about ages, like eternal truths about temporal distributional properties, are conceived to be truths *simpliciter* and fundamental. There are, on Cameron's view, by contrast no fundamental facts canonically expressed by means of the future- or past-tenses (Cameron 2015: 168). Presuming a suitable metric of time-units, and assuming that if *m* is of age *n*, then *n* is a number of time-units, we may accordingly lay down

(M_a) x is a concrete substance $\rightarrow \exists y(x$ is of age y & $\text{H}\neg(x$ is of age $y)$ & $\text{G}\neg(x$ is of age $y)$ & $\forall z(x$ is of age $z \rightarrow z = y))$

and add the following commentary:

$$m \text{ is of age } n \rightarrow \mathscr{T}(m \text{ is of age } n)$$

Cameron speaks of *concrete* substances to indicate that his focus is exclusively on things that exist in time; and he speaks of concrete *substances* to indicate that he is exclusively concerned with particulars (Cameron 2015: 7). Examples of concrete substances that Cameron provides include people, dinosaurs, and lunar colonies (*sic*) (*ibid.*).[5] We here take it as very natural to view the concrete substances that Cameron has in mind as residents of time in our sense of the term, although we must be careful not to generalize Cameron's claims to all residents of time, e.g. times or certain types of fusions. So far, so good. What are ages, though? Focusing on living things like us, the common conception of ages underwrites the following equivalence: m is of age n iff n time units ago, m was born (or conceived). Given (PER), the birth (or conception) of a given thing cannot, of course, be understood as the event of its coming into existence; but this finding does not as yet controvert the thought that, for living things, the aforementioned equivalence holds. However, since given (PER), I existed even before my birth and will exist even after my demise, and since there were then, and will then be, truths about how I am, was and will be, on Cameron's view, I must be said to have been of a certain age even at any time before my birth, and to be going to be of a certain age even at any time after my death.

This deviation from the common conception of ages need be no cause for concern, though. As long as there is an event of my birth, it is harmless to suppose that n units of time before my birth, I was of age $-n$, and that the counter keeps ticking even after my death, so that if I am presently of age n and my death occurs m units hence, k units after my death, I am of age $n + m + k$. Let us stipulate the following:

(D16) $\mathrm{D}_k\varphi \equiv_{\mathrm{df}} ((k < 0 \rightarrow -k$ units of time ago, $\varphi)$ & $(k = 0 \rightarrow \varphi)$ &
$(k > 0 \rightarrow k$ days hence, $\varphi))$

Accordingly, we can now replace the aforementioned equivalence by: m is of age n iff $\mathrm{D}_{-n}(m$ is born). Not all concrete substances are ever born, however, and yet, we are used to attributing ages to concrete substances like paintings or buildings or rock formations. Accordingly, the aforementioned equivalence must be taken to be restricted to living things. To generalise to all concrete substances, and keeping in mind that we here assume (PER), we may lay down: if m is a concrete substance, then m is of age n iff $\mathrm{D}_{-n}\exists x(x \text{ is present } \& \ m \mathrel{\mathrm{L}} x \ \& \ \forall y(y \prec x \rightarrow \neg(m \mathrel{\mathrm{L}} y)))$. In other

[5] Lunar colonies are not substances according to the traditional conception of what substances are, which suggests that what Cameron has in mind here really are concrete *particulars*. He also mentions the Scottish parliament in his discussion of entities in time that are in the scope of his version of the Moving Spotlight Theory (Cameron 2015: 209). Calling institutions 'substances' is equally non-standard, which confirms our suspicion. We will nonetheless follow Cameron's official terminology in presenting his view.

words, and assuming (PER), positive ages of a given concrete substance encode how long ago the time was present at which that concrete substance was first located; and negative ages encode how far hence the time will be present at which that concrete substance will first be located. Since I am not located at any time preceding my birth, this is still in line with the previous equivalence we considered, once the latter is restricted to concrete substances that are sometimes born.

Although, as we shall see in due course, this is not the end of the matter, let us pause to note that, even if the truth of an equivalence implies as yet nothing about any order of determination, it is hard to accept, as Cameron urges us to, that my present age determines how long ago the time was present at which I was first located, i.e. how long ago I was born. It is far more natural to contend – perhaps, even the only sensible thing to say – that I am of age n, if I am, because n units of time ago, I was born. Cameron, however, is adamant that the explanatory direction goes the other way, for he insists that, on his view, there are no fundamental facts whose canonical description would involve the past- or future-tense. *A fortiori*, on his view, said equivalence cannot be conceived of as providing a metaphysical explanation, or reduction, of what it is for a given thing to be of a certain age, since on that view, facts about ages *are* fundamental (Cameron 2015: 168).

As Cameron is well aware, however, it will not even do to say that if m is a concrete substance, then m is of age n iff $\mathrm{D}_{-n}\exists x(x \text{ is present} \mathbin{\&} m \mathrm{~L~} x \mathbin{\&} \forall y(y \prec x \rightarrow \neg(m \mathrm{~L~} y)))$ (Cameron 2015: 142–43). For, as long as it is not ruled that time has a beginning, there may be things that are *infinitely ancient*, where

(D17) m is infinitely ancient $\equiv_{\mathrm{df}}$ m is in time $\& \forall y(m \mathrm{~L~} y \rightarrow \exists z(m \mathrm{~L~} z \mathbin{\&} z \prec y))$

For such infinitely ancient things, there is no time at no earlier time than which they are located. And yet, even infinitely ancient things may have different properties as time goes by, and hence, on Cameron's account, such things must have ages.

Cameron accordingly suggest picking an arbitrary date such that a thing's age corresponds to how distant that date is from the time when the thing has that age. Of course, since precedence is not only rigid but eternal, 'how distant from' cannot here mean anything readily expressible in terms of (a metricised version of) precedence between times; for, otherwise, ages could not be said to change with time. So 'how distant from' must not here express any eternal relation, but is rather best understood in terms of how distant from the *present* time the chosen date is, or equivalently in terms of how distant from the time at which a given thing has a given age the *presentness* of the chosen date is. Accordingly, we may lay down that if m is a concrete substance, then m is of age n iff $\mathrm{D}_{-n}(d$ is present), where d is the arbitrarily chosen date.

One immediate consequence is that all concrete substances, infinitely ancient ones included, always share the same age, even if this age changes over time. This does not mean that the stages of the careers of these things go in tandem, in the sense that every two concrete substances ever born are born at the same age: the characterisation of the temporal distributional properties of these things will be made to match this arbitrarily chosen way to fix the numerical value of their ages.

The appeal to presentness is inessential, provided that there is at least one proposition that is only ever true once. Let us suppose that, for all ends and purposes, the proposition that Prior's heart stops beating qualifies. We can then use this proposition instead as a means to fix the numerical value of the postulated ages and accordingly say that a concrete substance *m* is of age *n* iff D_{-n}(Prior's heart stops beating). In this way, the current proposal escapes any troubles generated by invoking presentness.

The previously identified problem, however, now becomes even more prominent. On the suggested characterisation of ages, it is most natural to say that, in this sense of 'age', my age is determined by how long ago Prior went into cardiac arrest. After all, *in this sense* of 'age', I share the same age with every other concrete substance, including infinitely ancient things, and my age accordingly has nothing whatsoever to do with the time of my birth or with any other particularity of my existence or nature – but has rather everything to do with Prior's past death. According to Cameron, of course, facts about my age are *fundamental* and so are *not* determined by how long ago Prior went into cardiac arrest. Indeed, for Cameron, the latter is itself determined by Prior's present age, which Prior shares with me and any other concrete substance.

But his insistence to the contrary notwithstanding, Cameron's conception of ages nonetheless strongly suggests that, given the arbitrarily chosen date, my age – and that of all my permanently existing fellow substances – is determined by how long ago Prior went into cardiac arrest; and this would then mean that all the ways I am, was or will be ultimately depend upon Prior's past death – which is equally implausible. Either way, the emergent story lacks all plausibility. Cameron's conception of ages has lost its appeal, because it does not help to generate a convincing account of the way in which it is determined how concrete substances are, were and will be. The proponent of dynamic permanentism is therefore well-advised to abandon talk about ages altogether and, with it, (M_a). Instead, or so we now suggest, she should opt for the kind of view exemplified by a suitable instance of

(M) Ephemerally, *p*

where 'Ephemerally' is understood as follows:

(D18) Ephemerally, $\varphi \equiv_{df}$ sometimes, $(\varphi \mathbin{\&} H\neg\varphi \mathbin{\&} G\neg\varphi)$

We have assumed that the proposition expressed by 'Prior's heart stops beating' is ephemeral in this sense; but there are likely to be other candidates. For temporaryists, propositions expressed by sentences of the form '*m* is new' fit the bill. Dynamic permanentists will of course conjoin (M) with (PER); and so for them, such choices for '*p*' are foreclosed.

Once metric tense-operators are introduced, φ will be equivalent to '$D_0\varphi$', '$H\varphi$' will be equivalent to '$\forall k < 0\ D_k\varphi$', and '$G\varphi$' will be equivalent to '$\forall k > 0\ D_k\varphi$'.

Similarly, 'Sometimes, φ' will be equivalent to '$\exists k D_k \varphi$'. Consequently, given that, as a matter of tense logic, 'Sometimes, φ → Sometimes, Sometimes, φ' holds, (M) entails '$\exists k D_k p$'; and given the law of metric tense-logic

$$D_k\varphi \rightarrow D_{k'} (D_{k-k'} \varphi)$$

(M) ensures that always a different *k* is such that $D_k p$. To avoid any hostile takeover by the static permanentist, we can now add the following commentary:

$$D_k p \rightarrow \mathscr{T} D_k p$$

Consequently, against the backdrop of metric tense logic, proponents of (M) can account for the constant temporal change that dynamic theories demand.

Since the facts recorded by truths *simpliciter* need not be fundamental, a proponent of an instance of (M) is not *eo ipso* committed to there being fundamental facts canonically expressed by means of the future- or past-tenses. Equally, however, she may undertake to argue, on independent grounds, that Cameron's ban on such facts had better be rescinded in the end.

As we have argued, and bearing in mind the distinction between truths *simpliciter* and fundamental truths, Cameron himself is ultimately committed to some instance of (M). Which instance this is depends on his arbitrary choice of the date by reference to which it is fixed which *n* is such that we presently all are of age *n*. Likewise, the classical Moving Spotlight Theory and its Williamsonian counterpart can each be taken to imply an instance of (M); and as we have just seen, the same is true of presentism and GBT.

Unlike ($M1_c$), not all instances of (M) require there to be, for each time, instantaneous things located at it. Accordingly, (M) as such is ontologically less demanding than ($M1_c$). Moreover, unlike both ($M1_p$) and ($M1_c$), (M) as such avoids commitment to the intelligibility, or substantiveness, of either presentness or concreteness as applied to times.

Each instance of (M) implies (DYN) – at least provided that time has not both begun and come to an end, a possibility excluded by

(A14) PT ∨ FT, for T any chosen tautology

Combined with metric tense logic, instances of (M) thus guarantee that reality constantly changes and provide an answer to the question in what way it changes. (M) unfolds its explanatory potential only against the backdrop of metric tense logic which makes use of metric temporal operators such as 'D_k'. Although the framework devised in Chaps. 1, 2, and 3 does not include any principles governing metric operators, it can easily be expanded to encompass such principles.

We accordingly submit that dynamic permanentism is best understood as the conjunction of (PER) with some one instance of (M) – an instance that does not, however, involve the notions of presentness, concreteness or cognates, or Cameron's notion of ages. The instance of (M) generated by replacing 'p' by 'Prior's heart stops beating' will do.

This is not to say that dynamic permanentists endorsing such an instance of (M) cannot proceed to introduce a notion of presentness by means of the following partial definition:

(D19) $\mathrm{R}m \rightarrow (m \text{ is present} \equiv \exists x(m \mathrm{\ L\ } x \ \& \ \exists k(\mathrm{D}_k p \ \& \ \mathrm{At}\ x, \mathrm{D}_k p)))$

But once again notice that all the theoretical work is already done before this partial definition becomes available, and that the account it gives of the *definiendum* depends on the particular choice of 'p'. Accordingly, the introduction of 'is present' is a mere after-thought. The applicability of the notion as defined depends on whether the sentence replacing 'p' expresses an ephemeral proposition and on whether, always, for some k, '$\mathrm{D}_k p$' is a truth *simpliciter*; and it is certainly not built into any theory-independent notion of presentness, if such there be, that this is bound to be so.

We close this section by briefly examining which of the Barcan formulas and converse Barcan formulas dynamic permanentists accept, and which of them they reject. Since these formulas are perfectly general, and since even static permanentists accept that there are tensed sentences, the discussion can proceed at a level of generality at which the distinction between static and dynamic permanentism no longer matters. It is clear that, given (PER), the existential quantifier is always importable, and so, given (PER), all the converse Barcan formulas hold: whatever presently exists always exists. That ($\mathrm{BF_P}$) holds can be shown as follows: by classical quantification theory, $\mathrm{P}\varphi \rightarrow \exists x\mathrm{P}\varphi$ is a theorem, and hence by tense logic, so is $\varphi \rightarrow \mathrm{G}\exists x\mathrm{P}\varphi$. By quantification theory, then, $\exists x\varphi \rightarrow \mathrm{G}\exists x\mathrm{P}\varphi$ is also a theorem. By tense logic, then so is $\mathrm{P}\exists x\varphi \rightarrow \exists x\mathrm{P}\varphi$. The proof of ($\mathrm{BF_F}$) is similar. From these results, the Barcan formula for 'Sometimes' follows. In fact, from (PER) we not only get

$(\mathrm{BF_{At}})^{\forall}$ $\forall x((\mathrm{At}\ x, \exists y\varphi) \rightarrow \exists y \mathrm{\ At}\ x, \varphi)$ (with x and y distinct variables)

but also

$(\mathrm{BF_{At}})$ $(\mathrm{At}\ x, \exists y\varphi) \rightarrow \exists y \mathrm{\ At}\ x, \varphi$

Proof: Using (T2) and (T12), we derive '$(\mathrm{At}\ x, \exists y\varphi) \rightarrow$ Sometimes, (x is new & $\exists y\varphi$)'. Now since x and y are distinct variables, 'Sometimes, (x is new & $\exists y\varphi$)' entails 'Sometimes, $\exists y$(x is new & φ)', and so by ($\mathrm{BF_{Sometimes}}$) '$\exists y$ Sometimes, (x is new & φ)'. We can thus derive '$(\mathrm{At}\ x, \exists y\varphi) \rightarrow \exists y$ Sometimes, (x is new & φ)'. Using (T16) we can then derive '$(\mathrm{At}\ x, \exists y\varphi) \rightarrow \exists y \mathrm{\ At}\ x, \varphi$'. □

The proof, given in Chap. 4, to the effect that the Barcan formulas for 'H', 'G' and 'Always' fail, did not depend on any metaphysical view. Therefore, these formulas likewise fail given permanentism. Table 5.2 summarises these results and gives an

Table 5.2 Barcan and converse Barcan: permanentism, presentism and GBT

	Barcan Formula			**Converse Barcan Formula**		
	Permanentism	Pres	GBT	Permanentism	Pres	GBT
P	Yes	No	Yes	Yes	No	No
H	No	No	No	Yes	No	No
F	Yes	No	No	Yes	No	Yes
G	No	No	No	Yes	No	Yes
Sometimes	Yes	No	No	Yes	No	No
Always	No	No	No	Yes	No	No
At *m*	Yes	No	No	Yes	No	No
	$(BF_{At})^{\forall}$			**$(CBF_{At})^{\forall}$**		
	Permanentism	Pres	GBT	Permanentism	Pres	GBT
	Yes	Yes	Yes	Yes	No	No

overview of how the permanentists' commitments differ from those respectively incurred by presentists and proponents of GBT.

Against the background of these commitments, we can now establish a number of instructive results. Recall that the following was a theorem of the neutral framework developed in Chaps. 1, 2 and 3:

(T5) $\varphi \rightarrow \exists x \,\text{At}\, x, \varphi$ (x not free in φ)

Given that both GBT and permanentism are committed to (BF_P), on both these views a stronger claim than (T5) can now be established, *viz.*

(T19) $(\varphi \vee P\varphi) \rightarrow \exists x \,\text{At}\, x, \varphi$ (x not free in φ)

In other words, both GBT and permanentism entail that if something is or was the case, there is a time at which it is the case. To show this, we make use of (T5), (BF_P) and all of the following:

(A3) $H(\varphi \rightarrow \psi) \rightarrow (H\varphi \rightarrow H\psi)$
(R1) $\varphi / H\varphi$
(T3) (Sometimes, At x, φ) $\rightarrow$ (At x, φ)
(T5) $\varphi \rightarrow \exists x \,\text{At}\, x, \varphi$ (x not free in φ)

Proof: From (T5) we obtain, using (A3) and (R1), '$P\varphi \rightarrow P(\exists x \,\text{At}\, x, \varphi)$, from which, by ($BF_P$), '$P\varphi \rightarrow \exists x P(\text{At}\, x, \varphi)$' follows. By (T3), we have '$P(\text{At}\, x, \varphi) \rightarrow \text{At}\, x, \varphi$'. Using quantificational postulates, we derive '$\exists x P(\text{At}\, x, \varphi) \rightarrow \exists x \text{At}\, x, \varphi$'. We can thus derive '$P\varphi \rightarrow \exists x \,\text{At}\, x, \varphi$', from which (T19) follows with the help of (T5). □

Given that permanentism also validates (BF_F), a corresponding proof shows that permanentists are committed to the even stronger claim

$$(\text{Sometimes}, \varphi) \rightarrow \exists x \, \text{At } x, \varphi \quad (x \text{ not free in } \varphi)$$

Next recall that both proponents of GBT and presentists are committed to

(T12) $\text{T}x \rightarrow \text{At } x, x \text{ is new}$

This is of course a principle permanentists reject. We can now use (T12) and all of

(A2) $\varphi \rightarrow \text{GP}\varphi$
(A8) $(\text{At } x, \varphi) \rightarrow \text{T}x$
(T2) $(\text{At } x, \varphi) \rightarrow \text{Sometimes}, \varphi$

to prove the converse of (T19) – in other words, that if there is a time at which something is the case, it either is or was the case:

(T20) $(\exists x \, \text{At } x, \varphi) \rightarrow (\varphi \vee \text{P}\varphi) \quad (x \text{ not free in } \varphi)$

Proof: Let x be a variable that does not occur free in φ. By (T12), (A8), the logic of quantification and the logic of 'At', '$\text{E!}x \rightarrow (\text{At } x, \varphi \rightarrow \text{At } x, (x \text{ is new} \,\&\, \varphi))$' is a theorem. Consequently, by (T2) '$\text{E!}x \rightarrow (\text{At } x, \varphi \rightarrow \text{Sometimes}, (x \text{ is new} \,\&\, \varphi))$' is also a theorem. Given (A2) and the definition of 'x is new', '$\text{E!}x \rightarrow \neg\text{F}(x \text{ is new})$' is a theorem, and so it follows that the formula '$\text{E!}x \rightarrow (\text{At } x, \varphi \rightarrow ((x \text{ is new} \,\&\, \varphi) \vee \text{P}(x \text{ is new} \,\&\, \varphi)))$' – call it 'X' – is a theorem. But then '$\text{E!}x \rightarrow (\text{At } x, \varphi \rightarrow (\varphi \vee \text{P}\varphi))$' is also a theorem. Since x does not occur free in φ, by quantificational postulates it follows that '$(\exists x \, \text{At } x, \varphi) \rightarrow (\varphi \vee \text{P}\varphi)$' is a theorem. □

This proof can be modified to show that presentists can establish the stronger principle, namely

$$(\exists x \, \text{At } x, \varphi) \rightarrow \varphi \quad (x \text{ not free in } \varphi)$$

Indeed, as we saw in the proof of (BF_{At}) for presentism, '$\text{T}x \,\&\, \text{E!}x \rightarrow x \text{ is new}$' is a theorem of the presentist system. Hence, from the formula X of the previous proof one can infer '$\text{E!}x \rightarrow ((\text{At } x, \varphi) \rightarrow \varphi)$', from which we get the result.

Accordingly, GBT validates

(T21) $(\varphi \vee \text{P}\varphi) \leftrightarrow \exists x \, \text{At } x, \varphi \quad (x \text{ not free in } \varphi)$

By contrast, and given (T2) and a previously established result, permanentism validates

$$(\text{Sometimes}, \varphi) \leftrightarrow \exists x \, \text{At } x, \varphi \quad (x \text{ not free in } \varphi)$$

while presentism validates

$$\varphi \leftrightarrow \exists x \, \text{At } x, \varphi \quad (x \text{ not free in } \varphi)$$

These results chime with the intuitive thought that according to GBT, what is true at some time can be identified with what either is presently true or was true in the past, that according to presentism, what is true at some time coincides with what is presently true, and that according to permanentism, what is true at some time coincides with what is sometimes true.

This chapter has exclusively been concerned with the characterisation of GBT's main competitors, presentism and permanentism. Although it saw us criticizing certain extant proposals of how to formulate these views, the criticisms served merely as our starting points for devising better formulations on their proponents' behalf. No attempt has been made to argue that GBT is superior to these views as thus characterised. No such attempt will be made in the remainder of this book. Instead, we will concern ourselves with addressing certain objections that have been levelled against GBT in particular, against temporaryism, and against dynamic views in general. If these objections can successfully be answered, we will already have made considerable headway in the study of the subject. In the next chapter, we will accordingly address, and ultimately defuse, the perhaps most notorious objection against GBT, i.e. that it leads to an intolerable scepticism about our temporal whereabouts. This objection is fuelled by certain preconceptions of how GBT must construe the reality of the past – preconceptions which, as we shall demonstrate, rest on a confusion.

Chapter 6
The Epistemic Objection

Abstract In this chapter we critically discuss the so-called epistemic objection against the Growing Block Theory of time and argue that it rests on flawed conceptions of tense and of the import of the theory's main tenets. We show how the theory enables knowledge of the location of the edge of reality that it posits. After introducing the epistemic objection as it figures in the extant literature, we argue in Sect. 6.1 and Sect. 6.2 that this objection either rests on a gross misunderstanding of the theory's conception of the past, or else on a gross misunderstanding of the way in which utterances, or judgements, with tensed contents are evaluated for truth and falsity. In Sect. 6.3 we provide a constructive response to the remaining challenge, *viz.* to show how we might know that we are not in the past of the growing block's edge of becoming.

By far the most prominent objection to GBT is the so-called *epistemic objection*, first formulated by Bourne (2002) and further elaborated, in their distinctive ways, by Braddon-Mitchell (2004) and Merricks (2006).[1] Since then numerous authors have joined in the swan song for GBT (Heathwood 2005; Tallant 2007; Zimmerman 2011; Meyer 2013a; see also Braddon-Mitchell 2013). Our own view, that we undertake to substantiate in the following pages, is that the epistemic objection is ultimately the kind of philosophical 'howler' that Broad diagnosed McTaggart's famous argument to be (Broad 1938). Like McTaggart's argument, the epistemic objection, though seductive, ultimately founders because it does not take tense seriously enough.

Since we do not want to be accused of rigging the issue in our favour, the best we can do here is to quote at length from Bourne (2002), Braddon-Mitchell (2004) and Merricks (2006). We will then identify what we take to be major flaws in the reasoning that leads up to the objection. Once these flaws are exposed and rectified, we are left with nothing worth considering an objection to GBT, or so we shall argue (for the following see also Correia and Rosenkranz 2013 and 2015a).

Consider the challenge as first formulated by Craig Bourne who distinguishes between the ordinary, pretheoretical notions of being present and being past, on the

[1] The epistemic objection has a rather evident precursor in Lewis 1986: 93-94, which typically goes unacknowledged.

F. Correia, S. Rosenkranz, *Nothing To Come*, Synthese Library 395,
https://doi.org/10.1007/978-3-319-78704-6_6

one hand, and, on the other, GBT's notions of being located at the last time and being located only at times preceding the last time, for which latter he uses the expressions '*present*' and '*past*', respectively. Bourne writes:

> [T]he question [is] how we can know our time is *present*, for we would have all the same beliefs [...] even if we were *past* – Plato, after all, back in 365BCE, believes truly that he is teaching Aristotle, and it makes no difference to him that he is *past*! How are we not in the same position, according to no-futurism [i.e. GBT]? So here am I, a no-futurist, convinced that my present time is *present*. But wasn't I just as convinced yesterday, when I went through these arguments then? So, there am I as I was yesterday, as real as I am now, believing that I am *present*, and thinking pretty much the same things then about my previous selves as I think today. Yet I know now that my earlier self is mistaken; so how do I know that I now am not? (Bourne 2002: 362)

We pause to note that in suggesting that, at the time of writing, he knows his earlier self to be mistaken, Bourne either presupposes that, according to GBT, his earlier self is still believing to be *present* at that time – call this *presumption A* – or else presupposes that, while his earlier self no longer believes to be *present* at that time, according to GBT, its belief is nonetheless answerable to how things stand at that time – call this *presumption B*. For, if according to GBT, at the time of Bourne's writing, his earlier self no longer believes to be *present*, while its past belief was merely answerable to how things stood at the time it was entertained, there is evidently no basis for saying that, on a view such as GBT, Bourne's earlier self is mistaken.

Distinguishing between 'the current moment', i.e. the present time in the ordinary sense of the phrase, and 'the objective present', i.e. the last time in the sense GBT might be taken to give to that phrase, David Braddon-Mitchell likewise suggests that GBT has the following implications:

> A little over 2000 years ago, Caesar is crossing the Rubicon, believing he is doing so in the [objective] present. He is wrong. Of course once he was right: there was a time when that moment was the last moment of being, and then he was crossing the Rubicon in the [objective] present. But that time is gone. (Braddon-Mitchell 2004: 200–201)

Again, the reasoning would seem hostage to either presumption *A* or presumption *B*: Braddon-Mitchell's claim that, according to GBT, Caesar 'is wrong', but once 'was right', would lack any basis if, according to GBT, at the time of writing, Caesar no longer believed his crossing the Rubicon to be in the objective present, while his belief was merely answerable to how things stood when he was crossing the Rubicon a little over 2000 years ago. Braddon-Mitchell goes on to argue that the foregoing reasoning gives rise to an unacceptable scepticism:

> That then should lead us to wonder how we know that the current moment is in the [objective] present. From my current perspective I know that Caesar is in the objective past. But do I have any reason to believe that I am in the objective present? What if the objective present is in [a later year], when you, dear reader, are reading this paper. Here I am toiling in the past, to write something for you to read on the cutting edge of reality. Or perhaps the objective present is in 2017, or perhaps the universe is almost dead and the objective present is five billion years beyond [the current moment], and I am ancient history indeed. While we can tell that the objective present is not located in the past-directed volume of space-time from our perspective, there is no reason on the growing [block] view to think that the

> objective present is not located at any particular point in some volume of space-time that may lie in the future direction of us. Of course, if our current location [i.e. the current moment] is the objective present, then there is no future volume, but to know that our current location is the objective present we would need to know that there is no future-directed volume, and we have no independent access to this. So by a principle of indifference we should regard all alternatives as equally likely. So we should regard the hypothesis that the current moment is [objectively] present as only one among very many equally likely ones. So we should conclude, therefore, that the current moment is almost certainly in the [objective] past. This is absurd, and so by *reductio*, we should reject the growing [block] view. (Braddon-Mitchell 2004: 200–201)

Thus, according to Braddon-Mitchell, things are even worse for proponents of GBT than Bourne suggests, since they are committed to conclude, not only that we do not know that the growing block's edge is located now, but that we have overwhelming reason to think that it is located in the future of now. Note that the alternatives Braddon-Mitchell sketches would not be epistemic possibilities, and so by the principle of indifference, would not make it almost certain that we are in the objective past, if we could be said to *know* that we are in the objective present. Accordingly, the second stage of Braddon-Mitchell's reasoning crucially depends on the first, and so on either presumption *A* or *B*.

In a similar vein, and referring to Braddon-Mitchell (2004), Trenton Merricks writes that according to GBT,

> Nero is not (any longer) on the growing edge of being. So what are we to make of Nero's thoughts like 'I am sitting here at the present time'? The most obvious reply is that Nero is – and forevermore will be – thinking false thoughts, falsely thinking that he sits at the growing edge of being. [...] [But now] consider that you think 'I am reading this paper at the present time'. If 'the present time' refers to the growing edge of being, you ought to conclude that your own thought is false. After all, given [the theory of the] growing block, once you have a thought, you continue to have that thought forever. That thought is on the growing edge of being for just the briefest moment and is thereafter and forever not on the growing edge. As a result, the probability that your thought is on the growing edge is vanishingly small. Thus if Nero is wrong, then so – almost certainly – are you. That is an unwelcome result. (Merricks 2006: 105)

It would seem, then, that to avert this 'unwelcome result', proponents of GBT are obliged to give an account of how we might after all know that what is present, in the ordinary sense of the word, is on the edge of becoming. Merricks' reasoning here evidently relies on presumption *A*, i.e. the presumption that, according to GBT, Nero is still thinking that he sits on the edge of becoming. For it is this presumption that leads him to say that according to GBT, 'Nero is wrong'.[2]

The core of the epistemic objection to GBT accordingly is that, since in believing ourselves to be on the edge of reality, we are epistemically no better off than subjects who wrongly believe to be on the edge of reality simply because they are in the past of us, we can have no knowledge that, now, we are on the edge of reality.

---

[2] Admittedly, Merricks himself thinks that proponents of GBT should try to wriggle out of the impasse he here alleges by distinguishing between two notions of being present. However, this proposal proves to be a poisoned chalice. As we argue above, we have trouble seeing how the alleged problem arises in the first place.

Accordingly, for all we know, now is not on the edge of reality but in the past of it. Worse still, to the extent that we might wrongly believe to be on the edge of reality, just as subjects in the past of us do, it is far more likely, given our evidence, that we are not on the edge of reality, as there are then far more, equally likely occasions on which we believe wrongly. As argued, the stronger conclusion that we can be fairly certain that we are not on the edge of reality, depends on the weaker conclusion that we do not know that we are on the edge of reality; and the weaker conclusion crucially depends on the claim that subjects who are in the past of us wrongly believe to be on the edge of reality, or are now wrong in having believed this in the past, and *a fortiori* now lack that kind of knowledge. Neither conclusion would, all by itself, *refute* GBT. But, as we shall see in Sect. 6.3 below, it would imply that we can have no knowledge of GBT; and it is certainly odd to profess a theory that, *modulo* plausible auxiliary assumptions, implies its own unknowability. In what follows, we argue that both presumption *A* and presumption *B* are mistaken, which finding undermines both conclusions.

6.1 How Past Things Are

Merricks suggests that, given the truth of GBT, whenever once in the past, *x* was believing that she was on the edge of reality, then since *x* still exists, *x* is still believing that she is on the edge of reality. This is presumption *A*. Suppose then that, sometimes in the past, Nero was believing that he was on the edge of reality. Given the foregoing, it would follow that Nero is still believing that he is on the edge of reality; and evidently, the latter belief is false, because Nero is long since dead. But then, if Nero is long since dead, how can he currently be believing anything at all? If it really was a consequence of GBT that dead people are presently believing things, then it would hardly need a sceptical challenge to put it to rest!

What reason is there to charge GBT with such an absurdity? Describing the change that an event of point-sized duration undergoes when turning from being present to being past, Broad writes:

> When an event, which was present, becomes past, it does not change or lose any of the relations which it had before; it simply acquires in addition new relations which it could not have before, because the terms to which it now has these relations were then simply non-entities. [...] Nothing has happened to the present by becoming past except that fresh slices of existence have been added to the total history of the world. (Broad 1923: 66, cf. also 82)

Some commentators have – quite uncharitably in our view – interpreted the last sentence quoted as implying the hopeless thought that nothing ever loses any of its properties, tensed or untensed, by becoming past (Merricks 2006: 104–105; Zimmerman 2011; cf. also Sider 2011: 264). This would indeed have absurd consequences, given only that always, everything always in the future is something. For we would then be forced to hold, not only that C. D. Broad still exists, but also that, as we speak, he is still breathing, writing the first lines of *Scientific Thought*, etc.

If the past is as real as the present, and if this is now taken to imply that what went on at an earlier time *is still going on*, then indeed for all we can tell by inspecting what is going on around us, the edge of reality might lie in the future. For, we are then in no better epistemic position than someone located in the remote past who, on this uncharitable rendition of the view, might still be contemplating whether they are on the edge of reality, perceiving events that are still unfolding around them, although, alas and unbeknownst to them, reality has long since grown beyond any such event – a fact to which our own existence testifies. The image of the block would accordingly be that of a multi-storey building, with lower floors corresponding to the more distant past, where what happens on each floor is still happening, even if it is not happening on the last floor. But this evidently misconstrues the tensed metaphysics that GBT is meant to articulate: to say, on the one hand, that the past is real (exists), and hence that so are (do) the events that once occurred, is not to say, on the other, that past events are still occurring.

However, contrary to what authors like Merricks (2006: 104–105), Sider (2011: 264) and Zimmerman (2011) would seem to suggest, Broad's view does not imply the absurdity that always, for any Φ, tensed or untensed, everything that Φs is always going to Φ. To see this, assume that it is just past noon. If it is sufficient in order for any x last located at noon to have become past, that there be a new time succeeding noon – as Broad contends – then this should likewise suffice for such an x no longer to be Φ but to have been Φ instead, for any Φ such that while at noon, x is Φ, things can only ever be Φ at times at which they are located. Thus, let e be an event and let Φ be *occurs*, and assume (i) that at noon, e occurs, (ii) that $\forall x(e \text{ L } x \rightarrow x \leq \text{noon})$, and (iii) that $\forall x \forall y((\text{At } y, x \text{ occurs}) \rightarrow x \text{ L } y)$, where '$\leq$' is short for 'is earlier than or identical to'. If noon precedes now, this alone then ensures both that, at some time earlier than now, e occurs, and that now, e does not occur. Hence, e did occur but no longer occurs, once there exists a new time later than noon.

So, on Broad's view, in order for a present thing to have become past, indeed nothing needs to have happened to it 'except that fresh slices of existence have been added to the total history of the world'. For such an addition is quite sufficient in order for that thing to acquire past-tensed properties (if such there be) which it did not have before, and to lose present-tensed ones which it once had. As long as we allow as values for Φ properties that one possibly only temporarily has, such as the property of writing the first lines of *Scientific Thought* or the property of breathing, Broad can straightforwardly account for the fact that something that once had some such property now no longer has it (but instead has the property of having had it in the past, if there is any such past-directed property at all). That there are such temporary properties is something we are invited to presuppose, and indeed GBT is anyway committed to there being such properties, e.g. being new.[3] It simply does

[3]Assuming B-relations to be existence-entailing, Broad writes: 'When Queen Anne's death became, it came into relations with all that had already become, and to nothing else, because there was nothing else for it to be related to. All these relations it retains henceforth for ever. As more events become it acquires further relations, which it did not have, and could not have had while those events were non-existent. This is all that ever *happens to* the event in question. [...] All the

not follow from Broad's version of GBT, or ours for that matter, that if once Broad wrote the first lines of *Scientific Thought*, he is still doing so.[4]

These considerations also show that GBT has no need for Peter Forrest's 'highly controversial thesis' that the 'Past is Dead' according to which 'the hyperplane that is the objective present is the only one that contains consciousness' because 'consciousness is some by-product of the causal frisson that takes place on the borders of being and non-being' (Forrest 2004: 359; Forrest 2006: 162; the quotes are from Braddon-Mitchell 2004: 201). The fact that Broad once was, but no longer is, having any conscious thoughts is no more important than the fact that he once was, but no longer is, indenting the cushion of his office chair. It is unclear what further work Forrest's hypothesis is meant to do.

Accordingly, presumption *A* fails: according to GBT, even if Nero, who once believed to be on the edge of becoming, still exists at a time at which he no longer sits on the edge of becoming, it by no means follows that he is still believing, wrongly, to sit on the edge of becoming. The most that can be said about Nero in this respect is that he once believed to be on the edge of becoming, at which time he believed truly.[5]

---

relations which Anne's death entered into with the sum total of reality, as it was when this event first became, persist eternally for ever afterwards, and are wholly unaffected by anything else that may be added on to this sum total by further becoming. Hence no proposition *about these* will ever become false, and no false proposition *about them* will ever become true' (Broad 1923: 81–82; emphases added). Broad here merely talks about certain kinds of relations between events and about propositions concerning these relations. So nothing in these passages suggests that, according to Broad, all tensed propositions characterising a given event retain their truth-value, which would anyway be an odd thing for him to say, given only that on his view, once later events come into existence, what was present before no longer is present but past. As we have argued, on this view it is still accurate to say, as does Broad, that all that happens to a given instantaneous event *e*, in order for the proposition that *e* is present to become false and the proposition that *e* is past to become true, is that more events come into existence. What holds for the propositions ascribing to *e* the property of presentness (if such property exists) should likewise hold for propositions ascribing to *e* such temporary properties as *occurring*.

[4] Proponents of dynamic or A-theories of time often appeal to Arthur Prior's 'Thank goodness that's over' argument for tensed facts – an argument that was in fact anticipated by Broad although he never takes credit for this (Prior 1959; Broad 1938: 267, cf. also 527–33). In philosophical discussions, this argument is often given a reading that would not only tell against eternalism but also against GBT: to make sense of one's relief that one's pain is past, one has to say that the pain no longer exists; for if it did, it would still be hurting (see e.g. Zimmerman 2008: 215–16). Although we do not wish to rely on this argument, let us make two comments in response. First, like Prior himself but unlike Broad, friends of GBT may prefer not to admit events (or phases) such as pains into their ontology, in which case they might rather explain one's relief in terms of its no longer being the case that one is hurting – consistently with one's continued existence (Prior 2003: 7–19). Secondly, even if friends of GBT allow quantification over pains, they might still want to distinguish between the existence of an event and its unfolding, and reject the idea that insofar as the past pain still exists, it still is painful, just as we reject the idea that insofar WWI still exists, people are still dying in the trenches (*pace* Zimmerman 2008: 215–16).

[5] Cameron (2015: 64–65) discusses a principle which he calls *Past Record*:

> If something was the case, then it *is* the case in the past.

Some writers of a presentist persuasion, like Dean Zimmerman, have argued that thus understood, GBT has to reckon with 'ghostly' stuff, such as ex-parrots that lack any of the properties characteristic of parrots – a consequence they deem unpalatable (Zimmerman 2008: 215–16; 2011; see also Braddon-Mitchell 2013: 358). For what are ex-parrots if they are not parrots of some kind? Others, like for instance Timothy Williamson, have fully endorsed an ontology of such 'ghostly' things in order to sustain their permanentist views, but have argued that it is a bullet worth biting (Williamson 2013). Both camps would seem to presuppose, as uncontroversial, generalisations such as that always parrots are metabolizing, that always humans are breathing, that always dinosaurs are consuming space, etc.

This idea has an early precursor in Blake (1925). In his critical review of Broad (1923), Blake writes that 'the very nature of an event is to be an occurrence, a happening', which is why 'it is impossible for me to understand how an event can actually exist at a time when it is *not* happening' (Blake 1925: 427; emphasis in the original). Few, except perhaps certain presentists, would nowadays be inclined to follow Blake in his diagnosis concerning events. For, as Prior pointed out some time ago, one ought to distinguish – and in ordinary ways of thinking about the past, does distinguish – 'between the history that an event *has*, and the bit of history that it *is*' (Prior 2003: 10). Thus we frequently refer to past events, such as World War I, without conceiving of them as mere ex-events, rather than events proper, just because they are no longer in the process of unfolding. Similarly, we might very well argue that our classification of things into natural kinds – e.g. parrots, humans or dinosaurs – in no way requires that specimens so classified are still alive and kicking. It is of course a requirement that, say, anything that ever is a parrot sometimes throughout its existence metabolizes, but GBT can easily accommodate this thought, which is clearly weaker than the thought that always all parrots metabolize.

Once they adopt this weaker conception of what membership in a natural kind requires, it becomes clear that proponents of GBT have no need for 'ghostly' things of 'zombie' status, and thus can comfortably escape the challenge that Zimmerman and others articulate. They can thus also easily accommodate truths such as 'More humans have been killed by the Black Death than have died at the hands of the Spanish Inquisition' using simple set-theoretic notions. Unlike permanentists of the Williamsonian stripe, however, they can do so without being forced to give up on the natural enough thought, shared by other versions of permanentism, that humans

---

As long as 'in the past' is here taken to function as a temporal operator embedding a present-tensed clause, the principle merely records the way in which the past tense is regimented in the standard languages of tense logic. However, Cameron argues that Past Record is controversial and should ultimately be given up. His reasoning is based on an alternative reading of this principle according to which the italicized occurrence of 'is' is tenseless. It accordingly remains unclear what semantical role he assigns to 'in the past'; if the latter functioned like a temporal operator, it would be redundant provided that the 'is' of the embedded clause indeed is tenseless. But in any case, his contention – that, thus understood, Past Record should initially appeal to proponents of GBT – lacks all plausibility. It is clear as daylight that proponents of GBT will deny that if yesterday was the last day, then yesterday is atemporally the last day. The same goes for Nero once having believed to sit on the edge of reality.

are whenever they exist humans, tigers are whenever they exist tigers, etc. (cf. Williamson 2013: 8). This we take to be an advantage.

To conclude, even on Broad's own version of the view, GBT can combine the insight that by becoming past, things may lose some of their properties, and acquire others instead, with the insight that things continue to belong to the natural kinds to which our ordinary ways of thinking take them to belong. Thus, in claiming the past to be as real as the present, in that always everything always in the future is something, GBT is neither forced to obliterate the qualitative differences between what is present and what is past, nor is it committed to an ontology of 'ghostly' things that belong to kinds quite alien to those that our ordinary classifications aim to track. Moreover, unlike Williamson's version of permanentism, GBT can retain the insight that always every member of what we ordinarily think of as a natural kind belongs to that kind whenever it exists.

6.2 Looking Back Onto the Past

The upshot of the foregoing for a proper reconstruction of the epistemic objection is now this: One of the objection's central claims is that GBT implies that if Nero once believed to be on the edge of reality, he is now mistaken. But, as we have just argued, this claim cannot plausibly be understood to rely on the idea that, according to GBT, Nero still is believing to be on the edge. For, GBT clearly does not have this absurd consequence, its commitment to the continued existence of Nero, and of his relevant belief-episode, notwithstanding. But then, on what basis should it nonetheless be correct to say that, according to GBT, if Nero once believed to be on the edge of reality, he is now mistaken?

The underlying thought might be this – which brings us to presumption *B*. According to GBT, not only does Nero still exist, but so does his belief. Since the belief in question is a belief in the tensed proposition ⟨Nero is on the edge of reality⟩, and since this proposition is now false, so, it might now be suggested, is Nero's belief. More generally, the thought might be that individual utterances of tensed sentences, or individual judgements of, or beliefs in, tensed propositions, change their alethic status, whenever those sentences or propositions themselves do.

This is a very odd suggestion, though. Typically, we evaluate beliefs or judgements as true or as false according to whether their propositional contents are true or false at the time these beliefs and judgements are held or formed. For instance, if sitting in a long-distance train about to depart from Barcelona central station, Marta at that moment believes the present-tensed proposition ⟨The train is still in Barcelona⟩ to be true, her belief does not suddenly become false once the train crosses the city limits, just because the present-tensed proposition ⟨The train is still in Barcelona⟩ is then no longer true. The idea that individual judgements, beliefs or utterances, understood as historical occurrences, change their alethic status according to changes in worldly conditions, was rightly criticised by Gareth Evans (1985: 349–50). To the extent that they are attitudes towards present-tensed propositions, the only circumstances relevant for the evaluation of such historical occurrences would rather seem to be those that prevail when they are occurring. Accordingly, such token beliefs, token judgements and token utterances would seem to have a stable, rather than a variable, alethic status.

We may cast the latter idea in the form of a general semantic principle. Thus, let *S* be a declarative sentence-type, whether of a natural language or the language of thought; and let tokens of *S* accordingly include utterances if *S* belongs to natural language, and judgements and occurrent beliefs if *S* belongs to the language of thought, where in each case the content of the relevant utterances, judgements or beliefs is given by *S* in conjunction with features of the context of their occurrence. Then, plausibly, the following is a correct specification of the truth conditions for tokens of *S*, where the only token-reflexive expressions that *S* is assumed to contain are sensitive to time and to no other parameter:

(S1) x is a token of $S \rightarrow (x \text{ is true} \leftrightarrow \exists y(x \text{ L } y \text{ \& At } y, S \text{ is true}))$

We assume that tokens of sentence-types are located at specific times, which is also why, in case *S* belongs to the language of thought, we are exclusively concerned with datable judgements and occurrent beliefs, rather than, say, dispositions.

(S1) can be seen to involve a certain idealisation. For instance, since it takes time for it to be produced, a token-utterance of 'Now it is 9 o'clock sharp, and now it is a second after 9 o'clock' may intuitively be regarded as true on a given occasion although at no time does the sentence-type tokened on that occasion qualify as true. The idealisation then typically consists in the assumption that, at least for the purposes of giving a semantics for them, token-utterances can be treated as being instantaneous in the sense of

(D8) m is instantaneous $\equiv_{\text{df}}$ m is in time & Always, $\forall x$ Always, $\forall y(m \text{ L } x \text{ \& } m \text{ L } y \rightarrow x = y)$

(cf. Mulligan 2011 for the claim that token-judgements are best conceived along these lines). The idealisation is wholly on the part of semantics and has nothing in particular to do with GBT. (Note that the sentences relevant in what follows only contain one occurrence of the token-reflexive expression 'now'.)

As against this specification of truth-conditions for tokens, John MacFarlane has argued in recent years that, at least with respect to certain areas of discourse, utterance-truth is *assessment-context sensitive*: one and the same utterance, again understood as a historical act, may be true relative to one context of assessment and false relative to another, where metaphysically speaking, contexts of assessment are the same kind of beast as contexts of utterance, but unlike the latter make no contribution to the determination of what the utterance is about (MacFarlane 2003). Plausibly, what goes for utterances, goes for judgements and beliefs (see, however, MacFarlane 2003: 334, footnote 14). It might accordingly be suggested that while Nero's belief in the proposition ⟨Nero is on the edge of reality⟩ was true, as assessed from the time *t* at which this belief was formed, it is nonetheless false, as assessed from the present time, *even if we assume that Nero's belief concerns how matters stood at t.* MacFarlane's assessment-context relativism is a highly controversial doctrine and invites Evans' objection. Although MacFarlane has done a great deal of work in order to defuse the challenge Evans posed, we need not get into these subtle issues here. Suffice it to say that we want no part of assessment-context relativism. If the epistemic objection relies on a controversial doctrine such as MacFarlane's, then we can easily rid ourselves of any obligation to answer it by simply refusing to accept one of its controversial theoretical commitments.

With (S1) being in place, the question accordingly is whether Nero was already mistaken at the time when he came to believe the tensed proposition ⟨Nero is on the edge of reality⟩. If GBT holds, at that past time, that time was the last time, and so in particular, neither we nor now did as yet exist. According to (S1), in order to evaluate beliefs in tensed propositions formed in the past, we need to take into account how reality was back then; and it is an essential part of GBT that back then, reality had not yet grown sufficiently far so as to include either now or us. That reality has since then grown beyond that time, so that now that time is no longer on the edge, is immaterial for the assessment of Nero's past beliefs (cf. Button 2006: 132–33; 2007: 328–30). Consequently, presumption *B* likewise fails; and there is no longer any basis for contending, as champions of the epistemic objection do contend, that proponents of GBT must consider Nero's past belief as being currently mistaken. If Nero's belief was knowledge at the time this belief was entertained, it still is a piece of knowledge now, even if we cannot express the content of that knowledge in the very terms Nero is assumed to have used. In reporting what Nero knew, we must use the past tense. Yet, if Nero knew that he was on the edge of becoming, so we might know now that we are on the edge of becoming.

Of course, this is not to say that according to GBT we cannot, as of now, look back onto the past layers of the block, whose existence the theory itself affirms, and describe what was going on at those layers using the ontological resources available to us but not to our ancestors. But in so doing we must be careful not to unduly populate the reality which alone our ancestors' beliefs were answerable to, with things in time that came to exist only after those beliefs were being held. The manœuvre is familiar from the philosophy of modality. We can say, consistently, that there is a possible world in which the emperor Li Zhu does not exist, and so in which nothing is identical to Li Zhu, without thereby saying, inconsistently, that there is a possible world in which there is something, i.e. Li Zhu, such that nothing is identical to it. Similarly, in the temporal case. Just as 'Li Zhu' is a modally rigid designator of Li Zhu, 'now' is a temporally rigid designator of the present time. Accordingly, we can say that sometime in 1923, the event of *Scientific Thought*'s first being published was unfolding while now did not exist for another 95 years – and also say that for all times *t* later than 1923, sometime in 1923, the event of *Scientific Thought*'s first being published was unfolding while *t* did not yet exist – where all this is perfectly consistent with GBT's further claim that, at the relevant moment in 1923, there was no time such that the event of *Scientific Thought*'s first being published unfolded before it. And just as we can say that there is a possible world in which the Tang dynasty ends with Li Ye *and not Li Zhu* – something we could not have truly said about that world had it been actual – we can say that at some moment in 1923, it is that moment, *and not now*, that is on the edge of reality – something which GBT implies we could not have truly said about that moment at that very moment.

Similarly with respect to the time t at which Nero came to believe that he was on the edge of reality. According to (S1), Nero's belief is true iff, at t, 'Nero is on the edge of reality' is true. We are allowed to say that now there is a time t' later than t such that, at t, t' does not exist, all the while, at t, 'Nero is on the edge of reality' is true. But given how existence was defined, we are anyway not allowed to say that, at t, there is a time t' later than t that does not exist. If GBT is true, we are not allowed to say either that, at t, t' does exist – to say which would indeed be to imply that, at t, 'Nero is on the edge of reality' is false.

Could we still say that there now is a time t' such that, at t, $t < t'$? Yes, provided that, as we have assumed throughout, precedence is not existence-entailing. But even if precedence is not existence-entailing, we still cannot infer, from the claim that there now is a time t' such that, at t, $t < t'$, that at t, there is a time t' such that $t < t'$. For, as we have seen, on GBT, the converse Barcan Formula for operators of the form 'At m' fails. Thus,

$$\exists x \text{ At } y, \Phi x \rightarrow \text{At } y, \exists x \Phi x$$

does *not* hold, precisely because of some Φ which are not existence-entailing. Accordingly, either way we cannot, by simply exploiting the fact that there now is a time later than the time of Nero's belief, establish that Nero's belief was false when it was formed.

We must therefore conclude that there is no sense at all in which Nero is now mistaken, or was mistaken in the past when he formed his belief in the proposition ⟨Nero is on the edge of reality⟩, just because now the edge of reality lies elsewhere. But if there is no such sense, then the epistemic objection does not get off the ground. For the idea was precisely that since we are, epistemically speaking, no better off than Nero in any relevant respects, we do not now know that we are on the edge of reality, to the extent that Nero is (or was) mistaken in his corresponding belief.

This is arguably not enough by way of response, though. For we still owe a positive account of how we might know that we are on the edge of reality, and also how Nero might have known that he was on the edge of reality at the time he formed his belief. To this task we now turn.

6.3 Knowing to Be on the Edge of Reality

In the last section, we argued on independent grounds that the following semantic principle gives a correct specification of the truth conditions for sentence-tokens:

(S1) x is a token of $S \rightarrow (x \text{ is true} \leftrightarrow \exists y(x \text{ L } y \;\&\; \text{At } y, S \text{ is true}))$

where S is any sentence-type of natural language or the language of thought whose only token-reflexive expressions are sensitive to time and no other parameter.

We assume that competent speakers or thinkers have tacit knowledge of (S1) – knowledge that can be brought to consciousness through the kinds of reflections we engaged in before. We will now use this semantic principle in order to show how, given GBT, it follows that every token of 'Now is last' is true. On this basis, we will then argue that, given plausible principles of epistemic closure, subjects may know that they are on the edge of reality whenever they produce such a token, provided only that they know GBT. Finally, we will discuss and reject the suggestion that presupposing knowledge of GBT itself is somehow illegitimate in this context.

To begin with, recall the following two axioms that we laid down in the first chapter:

(A13) $(\text{At } x, \varphi) \rightarrow \text{Always, At } x, \varphi$
(A27) $x \text{ L } y \rightarrow \text{Always, } x \text{ L } y$

Let us now add the first tenet of GBT to the mix, i.e. the principle according to which never anything is ever going out of existence:

(P1) $\text{E!}x \rightarrow \text{GE!}x$

Given (P1), (A13) and (A27), (S1) already shows that it would be mistaken to assume, as some sceptics would seem to do, that for a proponent of GBT a token of 'Now is last' may be true sometimes in the past and then become false later on.[6]

Let us now consider the special case in which S is the type 'Now is Φ', for some Φ. Then given the indexical character of 'now', we have

(S2) $(\text{At } y, \text{'Now is } \Phi\text{' is true}) \leftrightarrow \text{At } y, y \text{ is } \Phi$

Taken together, (S1) and (S2) yield the following specification of the truth-conditions for tokens of 'Now is Φ':

(T22) $x \text{ is a token of 'Now is } \Phi\text{'} \rightarrow (x \text{ is true} \leftrightarrow \exists y(x \text{ L } y \,\&\, \text{At } y, y \text{ is } \Phi))$

Now let Φ be 'is last'. For this choice of Φ, (T22) yields

(T23) $x \text{ is a token of 'Now is last'} \rightarrow (x \text{ is true} \leftrightarrow \exists y(x \text{ L } y \,\&\, \text{At } y, y \text{ is last}))$

But now recall a theorem that we proved in Chap. 4:

[6] This also answers some of the worries that engage Evans (1985). We are at a loss to see, however, why this should imply, as Evans would seem to suggest it does, that tensed sentence-types cannot be true *simpliciter* without always being true.

(T13) $Tx \rightarrow$ At x, x is last

Given (T23) and (T13), and since tokens of sentence types are located at specific times, we can finally derive

(T24) x is a token of 'Now is last' $\rightarrow$ x is true

So in particular, presently any token of 'Now is last' is true; and hence we can produce such a true token, by laying down:

→ Now is last ←

(for the inwards pointing arrows that refer to the inscription they enclose, see Reichenbach 1947: 284). In the light of our reassurance that this token is true, the sceptical challenge can be considered as having been met head-on. Given your knowledge of (T24), you can make the very same move whenever you contemplate the matter, producing a token of 'Now is last' that you are in a position to know to be true whenever you produce it. The same applies to our ancestors, all the way down the family tree to Adam and Eve.

This optimistic conclusion depends on principles of epistemic closure, one's ability to recognise tokens as tokens of a particular sentence-type, one's knowledge of semantics, and most importantly, one's knowledge of GBT itself. While we may here be allowed to take the first three elements for granted, we have come across the objection that we cannot rely on the fourth (although Braddon-Mitchell, for one, would seem to argue that even if GBT is *known*, we cannot now know that we are now on the edge of reality; Braddon-Mitchell 2013: 352).

In the light of the aforementioned derivation, any threat to our knowledge that we are on the edge of reality would indeed be a threat to our knowledge of GBT – at least *modulo* epistemic closure, our ability to recognise the types to which tokens belong, and our knowledge of semantics. Accordingly, if there was such a threat, we could not brazenly assume that we do have knowledge of GBT. But as we argued in previous sections, the impression that there is such a threat is illusory.

If our knowledge of GBT *presupposed* our knowledge that we are on the edge of reality, then again, it would be problematical to appeal to our knowledge of GBT in order to show that we can have knowledge that we are on the edge of reality. However, this is so only to the extent to which knowledge of φ is not already said to presuppose knowledge of ψ, in the relevant sense of 'presuppose', whenever φ entails ψ. For then, we could never arrive at knowledge by inference without being charged with begging the question. So the question is whether in acquiring knowledge of GBT, we first need *to independently secure* knowledge that we are on the edge of reality. The relevant theorem of GBT was

(T13) $Tx \rightarrow$ At x, x is last

which, as shown in Chap. 4, depends on the two tenets of GBT

(P1) $E!x \rightarrow GE!x$
(P2) $Tx \rightarrow At\ x, H\neg E!x$

If in order to know (P1) and (P2), we would have to independently secure knowledge that we are on the edge of reality, then it would indeed be hard to see how we could know (P1) and (P2). The point can be made without relying on any reflections upon the epistemic situation of Nero and company. For, the notion of being last is a theoretical one; and it is correspondingly implausible to suppose that we can win through to knowledge essentially involving this notion without the help of any metaphysical theory that employs it.

But this very observation also casts doubt on the assumption that in order to know (P1) and (P2), we need to first know that we are on the edge of reality. Metaphysical reflection proceeds at the highest level of generality – given that all the principles of which the relevant theories are composed involve unrestricted quantification and can be prefixed by 'Always'. The methodological constraints with which the comparative assessment of such theories must comply, as well as the good- and the bad-making features of such theories, are likewise of a highly general character. It is therefore very unlikely that singling out one theory as better supported than any other would have to proceed by first establishing a particular thesis about us and our temporal vantage point – let alone one that involves notions whose proper application would seem to require metaphysical reflection.

From the onset, we declared that we would not here undertake the task to assess the relative merits and shortcomings of different theories of time and to show, on that basis, that GBT fares best overall. However, to avert the present charge, we do not need to argue that GBT is known. Given that we have already shown that if we were to know GBT, we would know that we are on the edge of reality, all we need to argue here is that our knowledge of GBT would not, in any problematical sense, have to piggy-back on independently acquired knowledge that we are on the edge of reality.

We conclude that the epistemic objection misfires because it presupposes a false claim – either the claim that past subjects now mistakenly believe to be on the edge of reality or the claim that, while they only used to believe they were, their past belief is retrospectively rendered false. We also conclude that we can know that we are on the edge of reality on the basis of knowing GBT – where our knowledge of GBT would not have to depend on independently secured knowledge that we are on the edge of reality. To this extent, the proponent of GBT has nothing to fear.

Chapter 7
Bivalence, Future Contingents and the Open Future

Abstract In this chapter we critically discuss the objection that since truths require grounds, the Growing Block Theory must take bivalence to fail for future contingents, while it proves at odds with the best account of such a failure. We challenge the version of the grounding requirement driving this objection, devise a better formulation, and show that the theory can retain bivalence and accommodate an interesting form of indeterminism. After rehearsing the objection in Sect. 7.1, in Sect. 7.2 we review different ways to articulate the grounding requirement, conclude that it should suffice that, for any tensed truth, sometimes there be grounds for it, and show how this requirement can be met by contingent truths about the future. In Sect. 7.3 we explicate a conception of the asymmetry between the open future and the fixed past, consistent with bivalence and available to the Growing Block Theory but none of its rivals.

In this chapter, we will formulate another challenge to GBT and then likewise defuse it. Throughout it is being presupposed that there are *future contingents*, i.e. statements about the future whose truth-value is not predetermined by the present or past – a presupposition that we are happy to make. The challenge then proceeds from a particular formulation of the grounding requirement on truth that would seem to force proponents of GBT to reject the principle of bivalence and to hold that future contingents are neither true nor false. It is then argued that the only sensible account of this failure of bivalence for future contingents – supervaluationism – is at odds with central tenets of GBT. We call this *the semantic objection* to GBT. Subsequently, we critically discuss the particular formulation of the grounding requirement on tensed truth that drives the objection, and propose an alternative, in our view much more appropriate formulation that proves to be compatible with both GBT and presentism. Thus, contrary to common conception, it so turns out that as far as the grounding of truth is concerned, the commitment to past things does not put GBT in any better position to account for the truth of statements about the past. Finally, we show to what extent GBT's acceptance of bivalence still allows for conceptions of the open future that are stronger than traditional forms of indeterminism and unavailable to permanentists: the future may be said to be open because time could come to an end, with no ontological commitment to future things standing in the way. While presentists can likewise opt for this conception of openness, they can

F. Correia, S. Rosenkranz, *Nothing To Come*, Synthese Library 395,
https://doi.org/10.1007/978-3-319-78704-6_7

only do so at the cost of jettisoning the asymmetry between the open future and the fixed past: for them, the past will be open in the corresponding sense in which time could just have begun, with no ontological commitment to past things standing in the way.

7.1 The Semantic Objection

Truths do not float free. They require grounds. This is easily seen, once we reflect upon the fact that saying something true is an achievement: whether our attempts to represent what the world is like are crowned with success depends on whether the world is the way we represent it as being. Accordingly, truths must be grounded in reality, because true representation is successful representation in this sense. The idea that truths must be grounded in order to qualify as truths is often put by saying that *all truths*, qua truths, *are grounded in what there is and how it is*. As we shall see in due course, this formulation must be taken with some care, as soon as we here understand the occurrences of 'are' and 'is' as being, without exception, in the present tense.

Consider a pair of future-tensed statements – respectively of the forms F*p* and F¬*p*, or the forms F_np and $F_n\neg p$, where *n* measures temporal distance and $F_n\varphi$ is understood in such a way as to entail Fφ – whose present truth-values are not already settled by how things located in the present or past of now are, or have been, in all their natural – non-future-directed, non-Cambridge-like – respects. For example, assume that both 'One day hence, some rain will fall' and 'One day hence, no rain will fall' belong to this category of statements. Statements of this category qualify as *future contingents* in that, even if one of them should presently be true, its present truth would anyway not be made inevitable, or be *predetermined*, by facts that are, strictly speaking, facts about what goes on in the present or what went on in the past.

That future contingents are not predetermined to be true in this sense does not *ipso facto* preclude that they are now both true and grounded. It is just that their truth, if they should be true, would not then be grounded in how things located in the present or past of now are or have been in all their natural, non-future-directed respects. Instantiations of Cambridge-like properties are unlikely candidates for being grounds for truth. Accordingly, if anything *now* grounds the present truth of 'One day hence, some rain will fall', it would either have to be something located in the present or past of now that has, or has had, some purely future-directed property – say, water-molecules, now hanging in a cloud, that presently have the historically contingent property of going to form raindrops one day hence – or something located entirely in the future of now like the event of tomorrow's rainfall. (Evidently, it is no good appealing to *causal* properties of things located in the present or past, apt to nomologically necessitate future facts and thereby to render the relevant statements already true now, since this would undermine the status of those statements as genuine future contingents.)

Suppose we follow Broad and assume an ontology of things in time exclusively composed of times and events. Then the first option is foreclosed, as no *event* located in the present or past of now is going to be a rainfall a day hence. It would clearly be cheating to say that the truth of 'One day hence, some rain will fall' is grounded in the fact that today's sunshine has the property of being such that, one day hence, some rain will fall: if anything, this is a Cambridge-like property of today's sunshine which can hardly be at the service of an account of *why* it is true to say that, one day hence, some rain will fall; and things are not getting any better, if we attribute such a property to the sum total of all events and times that are located in the present or past of now. Truth-grounding should not be said to be achieved by instantiations of such Cambridge-like properties of things located in the present or past.

Some philosophers have argued that *any* appeal to past- or future-directed properties – so-called Lucretian properties – amounts to cheating when it comes to showing that one's view complies with the grounding requirement on truth, never mind whether or not these properties are also Cambridge-like (Sider 2001: §2.3, Merricks 2007: 135). Thus, such philosophers likewise object to the idea that the truth of 'One day hence, some rain will fall' can be grounded in the fact that water-molecules that are now dispersed in a gaseous state presently have the property of going to form raindrops one day hence, or simply the fact that, one day hence, some rain will fall. The force of the objection against such tensed facts about the future is unclear. Thus, consider how an eternalist like Mellor (1998), who undertakes commitment to the existence of tenseless facts, will account for the truth of statements about the future such as 'One day hence, some rain will fall'. They will say that, if presently true, this statement is made presently true by the tenseless fact that it-is-raining-at-t, where t is some instant of tomorrow. This is not a case of cheating on anyone's count; and yet, the eternalist here invokes a tenseless fact that is, as it were, oriented towards the future. Why, then, are tensed theorists of time subject to the charge of cheating when they invoke the future-tensed fact that, one day hence, some rain will fall, to do the job? Being tensed, this fact does not always obtain, provided only that, on some days, it is sunny a day later. But why should this feature make one's appeal to the fact in question in any way objectionable in the current context?

All this is a matter of ongoing controversy, and we do not wish to pretend that the debate has been decided in anyone's favour (for a defence of Lucretianism, see Bigelow 1996). We here merely want to explore what follows from the assumption that Lucretianism is of no avail in this context. This would imply that the first option to account for the present truth of 'One day hence, some rain will fall' is indeed foreclosed, irrespective of whether one opts for an ontology of things in time exclusively composed of times and events.

However, according to GBT, neither does there presently *exist* anything that is located entirely in the future of now: tomorrow's rainfall does not now exist, and neither does tomorrow. So, the second option is likewise foreclosed. Consequently, there is, on this view, nothing at present such that facts about it could *presently* ground the truth of 'One day hence, some rain will fall'. This alone, surely, is not enough to establish that 'One day hence, no rain will fall' is presently true. Indeed, insofar as tomorrow no rain falls, some other event will then occur incompatible with its raining tomorrow; and *ex hypothesi* no such event presently exists, and *a fortiori* no facts about it which could ground the truth of 'One day hence, no rain will fall'. Were present truths to require present grounding, proponents of GBT would accordingly be committed to conclude that neither 'One day hence, some rain will fall' nor 'One day hence, no rain will fall' is true. Quite generally, where $F\varphi$ and $F\neg\varphi$ are future contingents, proponents of GBT would then be forced to regard neither statement as true.

Even if the truth of the negation of a given statement just is the falsity of that statement, it does not yet follow from this that these future contingents are neither true nor false: 'One day hence, no rain will fall' is not the negation of 'One day hence, some rain will fall'. Generally, $F\neg\varphi$ is not the negation of $F\varphi$, and so need not automatically be true whenever $F\varphi$ is false. What one would need in addition, in order to draw this further conclusion, is a principle like $\neg F\varphi \rightarrow F\neg\varphi$, or its metric counterpart, $\neg F_n\varphi \rightarrow F_n\neg\varphi$. These principles fail on so-called Peircean accounts according to which the future-tense operator is equivalent in meaning to 'Inevitably it will (in *n* units of time) be the case that'. For, if it is not inevitable that (one day hence) some rain will fall, this evidently does not entail that it is inevitable that (one day hence) no rain will fall. However, Peircean accounts are notoriously impoverished, as they leave us with no means at all to express the thought that something will be the case as a matter of mere historical contingency (cf. Prior 1967). In other words, we could not even formulate future contingents. In any case, this is not the understanding of the future-tense operator that our version of GBT, or Broad's version for that matter, presupposes.

Are we therefore bound to conclude that if all future contingents fail to be true, they likewise fail to be false? Not obviously so. The principles $\neg F\varphi \rightarrow F\neg\varphi$ and $\neg F_n\varphi \rightarrow F_n\neg\varphi$ are objectionable on other grounds. They in effect rule out that time has come to an end: if time has come to an end, then $\neg F\varphi$, $\neg F\neg\varphi$, $\neg F_n\varphi$ and $\neg F_n\neg\varphi$ should all hold. It is of no direct help to conditionalise these principles on the assumption that time goes on, for as long as it is deemed an open historical possibility that time has come to an end, this assumption will itself be a future contingent; and so if one initially thought that all future contingents are uniformly false, one will be able to accept the conditionalised principles already because one takes them to have a false antecedent. Unlike the unconditionalised principles, the conditionalised principles would accordingly seem to put no pressure on those who try to resist the conclusion that, just because they are untrue, future contingents are neither true nor false.

However, even if one cannot rule out, on purely logical grounds, that time has come to an end, one may still accept $\neg F\varphi \rightarrow F\neg\varphi$ and $\neg F_n\varphi \rightarrow F_n\neg\varphi$, or relevant instances of them, as historically necessary truths, because one thinks that something about the present and past makes it inevitable that time will go on indefinitely, in whatever way, or will at least go on for another day. In that case, one is indeed driven towards the conclusion that if neither 'One day hence, some rain will fall' nor 'One day hence, no rain will fall' is true, neither of them is false either.

That statements like 'One day hence, some rain will fall' and 'One day hence, no rain will fall' are neither true nor false is a conclusion that Broad himself explicitly draws. He contends that while present-tensed statements about the non-existent such as 'Puck exists' are uniformly false, statements about the future are without exception neither true nor false (Broad 1923: 70–73). His reason for such unequal treatment is as follows. According to Broad, those who assert the statement 'Puck exists' thereby assert 'that some part of the existent has [the] set of characteristics' 'by which they [describe] Puck to themselves', where this is rendered false by the negative fact that nothing exists that has those characteristics (Broad 1923: 71). By contrast, Broad maintains that the statement 'Tomorrow will be wet' can only be rendered true or false by the fact of tomorrow's being wet or the fact of tomorrow's being dry 'when tomorrow comes'; and since tomorrow has not yet come, there is presently no fact that could render this statement either true or false (Broad 1923: 73).

If, in the statement 'Tomorrow will be wet', 'tomorrow' was meant to function as a singular term or a definite description taking wide scope over the future tense, then it is unobvious why Broad should not have said, instead, that the statement is false. After all, if anything, tomorrow is the day after today; and according to GBT, there is at present no such day. Thus, it seems a much more charitable interpretation to assume that, in Broad's statement 'Tomorrow will be wet', the future tense takes wide scope over 'tomorrow', and that a more perspicuous, tense-logical rendition of that statement would be 'It will be the case that the day after today is wet', which, being of the form $F\varphi$, is not relevantly different from our earlier 'One day hence, some rain will fall'. Indeed, at a later stage of his discussion, Broad himself claims that what one *asserts* in making a statement such as 'Tomorrow will be wet' is 'that the characteristic in question [here: being wet] will characterise some part of what will become [here: what will then be the next day]', to assert which, Broad adds, is 'compatible with the non-existence of the future' (Broad 1923: 77).

Although we ultimately agree with Broad's observation that the truth of such a future-tensed statement depends on how, in the future, matters will stand, there are at least two reasons for being dissatisfied with his narrative. First, if what is asserted by means of such a statement is that the characteristic in question *will* characterise something, why should the observation that nothing has as yet come to exist that has, or is going to have, that characteristic, in any way suggest that the statement is not yet either true or false?

Secondly, there is something singularly odd about Broad's suggestion that a present statement about the future such as 'It will be the case that the day after today is wet' will be true or false only once tomorrow arrives.[1] What is certainly congenial to the tensed metaphysics that GBT provides is that one day hence, the present-tensed statement 'The present day is wet' will be rendered true or false by some fact, or event, that will not have rendered it true or false any earlier. If, as Broad

[1] This idea has been elaborated, in their respective ways, by Belnap et al. (2001) and MacFarlane (2003).

himself would seem to suggest, however qualifiedly so, always whatever is the case was going to be the case (Broad 1923: 78), the future truth of 'The present day is wet' one day hence should imply today's truth of 'It will be the case that the day after today is wet'. At least, there is no reason to deny this truth-value link, if there is no antecedent reason to claim that the statement presently lacks a truth-value. In any case, however, one would have expected that whatever the present truth of 'It will be the case that the day after today is wet' presently requires, *it is not the present existence of the kind of fact that a day hence, the truth of 'The present day is wet' will require*. Yet, if this is so, there is again no evident reason for thinking that just because that fact does not now exist, the future-tensed statement is not now true.

Broad would seem to draw a principled distinction between what the truth of a statement about the future would require – according to him, that there *is* a kind of fact that does not yet exist but will only ever exist in the future – and what it asserts – that there *will be* a fact of that kind. Yet, such a distinction should not impress as very natural and certainly requires argument. But even if it can be drawn, this does not render Broad's particular way of drawing it the least plausible: why should the truth of 'In the future, *p*' require the existence of the fact that *p*? If the fact that *p* presently existed, then plausibly, '*p*' would be true. Again, why should we be tempted to concede that the truth of 'In the future, *p*' is only ever to be had at the cost of '*p*' being already true? As will become clear in due course, similar sorts of considerations are suited to disarm the thought, sketched earlier, that insofar as GBT holds, future contingents fail the grounding requirement on truth, properly construed.

If one accepts, as Broad does, that future contingents present counterexamples to the principle of bivalence, one is accordingly faced with the question of what logic and semantics one ought to assume when reasoning about the open future. On this matter Broad remains frustratingly silent. The first option that comes to mind is to model truth-value gaps along the lines suggested by Kleene's or Łukasiewicz' three-valued logics. But these choices have well-known drawbacks. Even if bivalence fails because it is as yet open whether or not sometimes in the future, some rain falls, it is nonetheless desirable that $F\varphi \rightarrow F\varphi$ be valid. Yet, in Kleene's (strong) three-valued logic, if $F\varphi$ is neither true nor false, so is $F\varphi \rightarrow F\varphi$. On Łukasiewicz' logic, no comparable problem arises, because $F\varphi \rightarrow F\varphi$ will be true, even if $F\varphi$ is neither true nor false. However, on either choice of logic, $F\varphi \vee \neg F\varphi$ will be neither true nor false when $F\varphi$ is a future contingent, and so will be $\neg(F\varphi \,\&\, \neg F\varphi)$; and this is clearly undesirable: even if it may presently be unsettled whether or not some rain will fall, it should anyway be settled that either it is the case that some rain will fall, or it is not the case that some rain will fall, and that anyway not both is the case (cf. Prior 1953). It is for these reasons that *supervaluationism* seems to be a much better choice, in order to model the denial of bivalence for future contingents which, given the foregoing, proponents of GBT would seem to be committed to (Thomason 1970; cf. also MacFarlane 2003). For, the non-bivalent semantics that supervaluationism affords still underwrites all theorems of classical logic.

Supervaluationism assumes that time is forward-branching but not backward-branching. Thus, where we take the variables '*u*', '*v*' and '*w*' to range over times, supervaluationism implies (i), which rules out backward branching, but not (ii), which, if it held, would rule out forward branching:

(i) $\forall u \forall v \forall w(v \prec u \;\&\; w \prec u \rightarrow (v \neq w \rightarrow v \prec w \vee w \prec v))$
(ii) $\forall u \forall v \forall w(u \prec v \;\&\; u \prec w \rightarrow (v \neq w \rightarrow v \prec w \vee w \prec v))$

According to supervaluationism, φ is true at a time *t* just in case φ is supertrue at *t*; and φ is false at time *t* just in case φ is superfalse at *t*. Supertruth, and superfalsity, at a given time *t* are defined as follows: φ is supertrue at time *t* just in case φ is true at *t* on all histories that include *t*; and φ is superfalse at time *t* just in case φ is false at *t* on all histories that include *t*. Here, a history is a maximal linear set of times, where a set *s* of times is linear iff $\forall u \forall v(u \in s \;\&\; v \in s \rightarrow (u \neq v \rightarrow u \prec v \vee v \prec u))$, and is a maximal such set iff there is no set of times s^* such that s^* is linear and *s* is strictly included in s^*. For all histories *h* and any $t \in h$, truth at $\langle t, h\rangle$ behaves classically; and Fφ is true at $\langle t, h\rangle$ iff there is a $t^* \in h$ such that $t \prec t^*$ and φ is true at $\langle t^*, h\rangle$. Accordingly, given how supertruth was just defined, for any φ, whether or not it is a future contingent, both φ ∨ ¬φ and ¬(φ & ¬φ) come out supertrue at any time, even if there are times at which neither φ nor ¬φ is supertrue. Similarly, Fφ → Fφ comes out supertrue at any time, even if there are times at which neither Fφ nor F¬φ is supertrue.

However, it should be clear that the supervaluationist treatment cannot be made to square with one of the consequences of GBT, *viz.* that there is no time later than now. For if now is the last time, there can be only one history, that history having a last moment, namely now. Accordingly, given the supervaluationist's semantic treatment of the future tense operator F and of the operators ¬ and &, *all* statements of type ¬Fφ & ¬F¬φ must be taken to be supertrue now, and hence to be true now. This means that given the supervaluationist view, GBT is committed to time having just come to an end. This is, of course, an unfortunate result.

Here then is, at last, the semantic objection to GBT: insofar as present truths need to be presently grounded in what there presently is and how it presently is, where the characterisation of how things presently are must make no appeal to any Cambridge-like or future-directed Lucretian properties, proponents of GBT are bound to treat future contingents as being neither true nor false, lest they be committed to assuming that time will not go on. However, the best way to accommodate the thought that future contingents are neither true nor false is to opt for supervaluationism: all other attempts to model truth-value gaps have unpalatable consequences, including rejection of the law of excluded middle. Yet, supervaluationism is ultimately at odds with central tenets of GBT.

7.2 The Grounding Requirement on Truth Revisited

The semantic objection to GBT heavily relies upon a particular formulation of the grounding requirement on truth, and tensed truths in particular. This formulation is not the only one we might give, though. Thus, there is a deflationary take on the grounding requirement that exploits Aristotle's insight that '[i]t is not because we think truly that you are pale, that you are pale, but because you are pale we who say this have the truth'.[2] The core idea of such a deflationary approach is that we can straightforwardly discharge the grounding requirement, properly understood, by simply using the statement, or articulating the proposition, whose truth is in question, in order to state what grounds its truth (Tallant and Ingram 2015). Suitably generalising from Aristotle's dictum, we arrive at a schematic principle according to which for any substitution instance for '*p*', '*p*' is true, if it is true, because *p* – where it is here being assumed as part of the logic of 'because' that claims having it as main connective behave asymmetrically: if 'ψ because φ' holds, then it is not the case that 'φ because ψ' holds (see Correia and Schnieder 2012: 8, 26).

According to this deflationary approach, that truths be grounded amounts to no more than that suitable instances of the Aristotelian schema should hold for the truths in question; and that these instances hold is taken to follow from an explication of the notion of truth itself. There is then no need to come up with any more informative specifications of the grounds for truths – let alone any specifications that relate to what presently exists, unless the truths in question are themselves existential claims. Thus, for the case of future contingents like 'One day hence, some rain will fall', this means that there is as yet no pressure at all to explain why they are true, if they are true, by appeal to the kinds of purely futurely located things whose existence GBT denies.

Although we have sympathies for this deflationary approach to truth-grounding, we recognize that it is likely to leave objectors with the feeling of having been short-changed. After all, the question was precisely what in the reality of things and their properties and relations that alone GBT acknowledges to exist – i.e. in the block grown as far as the present – can possibly account for the truth of future contingents; and this question does not go away by helping oneself to those future contingents in explaining their truth *on condition that they are true at all*. Rather, this question immediately translates into one about whether that sort of explanation can ever be accurate, because there may just turn out to be nothing that would ever ground discharge of the assumption that such future contingents *are* true.

Let us therefore concede that more needs to be said about truth-grounding than the deflationary approach supplies. Let us also concede to the objector that truth-grounding is always in terms of what there is and how it is. These concessions are not yet an admission of defeat in one's attempt to reconcile GBT with the truth of future contingents. For as long as no case has been made that, in general, in order for a statement to be presently true, it must *presently* be grounded in what there is

[2] *Metaphysics*, 1051b6–8; the translation follows Ross et al. (1908).

and how it is, GBT has nothing to fear; and it is precisely such a case that we think just is not forthcoming.

To begin with, let us ask again: Why should the truth of a given statement require grounding by something that the statement itself does not even claim or assert to exist? According to GBT, a statement of the form 'A thousand years hence, a child will be born' – unlike a statement of the form 'Some child will be born a thousand years hence' or 'Something will a thousand years hence be a newly born child' – does not affirm the existence of anything. How do we explain this difference in content between these different types of statements, if we simultaneously maintain that the present truth of all of them alike requires the present existence of something that will a thousand years hence be a newly born child?

It is admittedly unclear what the force of these intuitive considerations is. Arguably, many truths are grounded in things that they are not, in any obvious sense, 'about' (*pace* Merricks 2007). Thus, on certain views, statements such as 'Putin is reckless' require the existence of tropes – e.g. Putin's particularized recklessness – in order to qualify as true, while it is certainly too far a stretch to suggest that these statements thereby affirm or are 'about' the existence of such tropes. And yet, when on a view such as GBT, statements of the form '$F\exists x\Phi x$' and '$\exists xF\Phi x$' receive a systematically different semantic treatment, then this has its basis in the thought that prefixing an existential statement by a future-tense operator is a means to cancel any ontological commitment – not only to anything that presently Φs, but also to anything that futurely Φs. There certainly are operators suited to effect such a cancellation (e.g. 'John believes that'); and until GBT's claim that future-tense operators belong to this broad kind has successfully been dislodged, insistence that truths of the form '$F\exists x\Phi x$' nonetheless require the existence of something satisfying the open sentence '$F\Phi x$' is dialectically ineffective.

Even if there was after all presently a thing of which it was true to say that it will be a newly born child a thousand years hence, its mere present existence would anyway fail to ensure that this is what can at present truly be said about it. This is why the grounding of truth is said to be in terms both of what there is and *how* what there is *is*. But now, even if Émilien presently exists, and according to GBT will continue to exist forever, why should how he presently is ground the truth of the future contingent 'Seventeen years hence, Émilien will have a ball'? That Émilien will have a ball in seventeen years' time is, plausibly, not prefigured in any of his currently exemplified properties in conjunction with his present existence – unless, that is, we allow Lucretian properties among those properties.

Consequently, its commitment to past things does not automatically put GBT in a better position to account for the grounding of truths about the past: that Nero, located in the past of now, still exists, is insufficient to ground the present truth of 'Nero was mad'. The task might become easier if one pairs GBT with an ontology of tropes or events, for arguably Nero's particularized madness is sufficient to ground the truth of 'Nero was mad'. But as long as GBT is intended to be open to those metaphysicians who reject ontological commitment to tropes or events, it is as yet no better placed to account for the truth of statements about the past than to account for the truth of statements about the future. The thought generalises to other views: if Lucretian properties are out of bounds, and if all residents of time other than times are 3D objects that persist by enduring, even permanentists of the Williamsonian stripe face problems when asked to provide present grounds for truths about the past or future.

Suppose we continue to disallow Lucretian properties of any kind. Then there is no more mystery in the thought that the present truth of 'F(something Φs)' *will in the future be* grounded in something that then Φs, than in the thought that it is partially grounded in the present existence of something of which it is presently true to say that it will Φ, where only in the future, the truth of the latter predication will be grounded by that thing's Φ-ing. Indeed, why should it not in general be enough in order for a statement about the future to be presently true, that its present truth will in the future be grounded by something being certain ways, whose future existence and future ways of being the statement now affirms – just as it is enough in order for a statement about what happens at a distance to be true here, that at a distance, its being true here is grounded by something's happening whose distant existence and happening the statement here affirms?

Suitably generalised, this view holds that the truth of a given tensed statement at most requires that it *sometimes* be grounded in what then is something and a certain way, provided that the statement claims that there then is such a thing that then is that way (Westphal 2006; cf. also for further discussion Gallois 2004; Kierland and Monton 2007; Baia 2012; and Tallant and Ingram 2015).[3] This theoretical option shows that taking tense to be irreducible need not automatically commit one to the Lucretian answer to the truth-grounding problem, contrary to what Sider (2001: 39) concludes.

In particular then, the present truth of a statement about how, at some future time, things will be, might well be said to be, at that future time, going to be grounded by things being that way. But how, one might ask, can such future grounding *ensure the*

[3] Dummett argues that 'a proposition about what I am going to do is true in virtue of my later action' and, more generally, that '[i]t is *what is going to happen in the future* that renders our statements about the future true, when they are true. This platitude is embodied in the truth-value links' (Dummett 2004: 81, 83). Dummett concludes that a view like Broad's, according to which there is now nothing that is only ever going to happen in the future, is bound to deny that statements about the future are true (Dummett 2004: 74, 80). Contrary to what Dummett claims, however, the truth-value links merely yield that a statement about the future such as 'It will be the case that Mars is being colonized' is presently true just if it will be the case that 'Mars is being colonized' is true. It is only against the backdrop of Dummett's further contention that 'a proposition can be true only if there is something in virtue of which it is true' (Dummett 2004: 74), that we can reason from this to the conclusion that if 'It will be the case that Mars is being colonized' is presently true, there is now something in virtue of which 'Mars is being colonized' will be true, i.e. a merely future happening. The present suggestion is to replace Dummett's contention by the claim that a proposition is true at t only if *sometimes* there is something in virtue of which it is true at t. Note also that, contrary to what Dummett (2004: 80) suggests, the proponent of GBT need not treat the truth-conditions of statements about the past in any substantially different way – irrespective of the fact that, according to GBT, past things still exist (cf. also Broad 1938: 316).

present truth of such a statement? The answer to this question is pretty straightforward.

Our toy example will again be the future contingent 'One day hence, some rain will fall', which we assume to be presently true. For sake of perspicuousness, we use 'q' as shorthand for the embedded clause 'Some rain falls', '$\downarrow$' for 'One day ago', and '$\uparrow$' for 'One day hence'. The future contingent in question can accordingly be written as '$\uparrow q$'; and the task before us correspondingly is to explain why '$\uparrow q$' is presently true. Our explanation relies on two general principles, meant to be restricted to those φ which are grounding-theoretically unproblematic and naturally thought of as being in the simple present-tense:

(E1) Always, if φ is true, then $\exists X$(φ is true because X exist)
(E2) Always, if φ is true, then $\downarrow$('$\uparrow\varphi$' is true) because φ is true

The plural quantifier in the first principle (E1) ranges over pluralities of the kinds of entities – things, facts, tropes, events, *etc.* – that according to one's preferred ontology, conspire to make the unproblematic statements true when they are true. Thus understood, (E1) can be assumed as common ground. So can (E2): this second principle encodes the natural enough thought that whenever a given unproblematic statement φ is true, '$\uparrow\varphi$' was true a day ago and its having been true a day ago is grounded in the truth of φ. By the transitivity of 'because' – another plank of the common ground – (E1) and (E2) together yield:

(E3) Always, if φ is true, then $\exists X$($\downarrow$('$\uparrow\varphi$' is true) because X exist)

With (E3) in place, we can now reason as follows:

(1) '$\uparrow q$' is true
(2) $\uparrow$('q' is true)
(3) $\uparrow\exists X$($\downarrow$('$\uparrow q$' is true) because X exist)

From our initial assumption (1), we derive (2) using Tarskian conditionals; (2) and (E3) together then yield (3) which may be taken to say the same as

(4) One day hence, there will be things whose existence will explain <u>why one day before, '$\uparrow q$' was true</u>

But reflecting on what the underlined clause in effect says, it is easy to see that (4) is tantamount to

(5) One day hence, there will be things whose existence will explain <u>why '$\uparrow q$' is true now</u>

where 'now' takes us right back to the present time. (5) thus provides us with a successful grounding explanation of why '$\uparrow q$' is presently true.

We are now also in a position to formulate a general grounding requirement that does justice to the idea that truths do not 'float free', goes beyond a simple deflationism and, at the same time, is hospitable to tensed ontologies. Thus, let ψ range over present truths of the problematic kinds – including contingent truths about the future – then we can lay down:

(GR) For all ψ, Sometimes, $\exists X$((Now, ψ is true) because X exist)

where, as before, 'Now' is a temporal operator that always shifts the time of evaluation back to the time of utterance.

(GR) allows for the possibility that a statement about the future be true now whose present truth is sometimes in the future explained by what exists then, while there exists now nothing that would explain its present truth. However it might now be suggested that this is not yet to allow for the truth of future *contingents*. Thus, it might be suggested that if a statement about the future like 'One day hence, some rain will fall' is indeed presently true, *that alone* is guarantee enough to conclude that the future is bound to be a certain way – eliminating thereby the possibility of truth for future *contingents*. In reply, it should first be noted that if we could, in principle, presently know for certain that some statement about the future is true, then given our principled inability to inspect the contingent future course of events, this would indeed suggest that that statement is not a future contingent. But such epistemic considerations are of no present concern. Rather, what is in question here is whether the present truth of a statement that will be grounded by how matters are going to stand, in any way undermines its status as a future contingent. An affirmative answer to this question, we submit, gets things backwards.

If we already built it into the very notion of a future contingent that no statement about the future can qualify as a future contingent *if its truth is already entailed by all present truths* (cf. Markosian 1995: 96), then it would trivially follow that future contingents are never true, and proponents of GBT would be ill-advised to even try to reconcile their metaphysical views with the truth of future contingents. This would still, however, allow formulation of the sensible question of whether there are any truths about the future that are not made inevitable by any present or past *things, facts, states or happenings*. Present truths about the future that will be grounded by what there will be and how it will be, may well qualify as such, as long as nothing there is or was, in conjunction with how it is or was, makes it inevitable that, in the future, there will be such grounds (Broad 1937: 204, 206; cf. also Rosenkranz 2012).[4] This is the more natural – and more neutral – conception of future contingents that leaves the question of their truth as yet open. In any case, it is this concep-

[4] Later Broad clearly distinguished between the future's being predetermined by past and present facts and its being 'predeterminate' in the sense that statements about the future have a definite truth-value (Broad 1937: 204, 206). But since he would seem to have abandoned his 1923 view by that time, he never readdressed the question of why, in the light of this distinction, GBT should be taken to be committed to denying that future contingents are ever true.

tion that we here assume when arguing for the conclusion that GBT is compatible with the truth of future contingents; and once the grounding requirement is properly construed as demanding no more than that every truth sometimes be grounded in what there then is and how it then is, such compatibility is indeed hard to deny.

Admittedly, we so far have not said enough about how the truth of the negation of a future contingent is ever grounded, and so, given that the falsity of a statement consists in the truth of its negation, we have not said enough either about how the general principle of bivalence can be reconciled with a sensibly temporalised grounding requirement on tensed truths.

We must distinguish between different cases. Consider first future contingents of the form $F_n\varphi$. Granted that time will go on for at least n units of time, the truth of $\neg F_n\varphi$ can be reduced to the truth of $F_n\neg\varphi$, whose grounding can in turn be explained along the aforementioned lines. If, by contrast, time will *not* go on for at least n units of time, then one can hold that for some m, with $0 \leq m < n$, $\neg F_n\varphi$ is true because m time-units from the present, time has come to an end, and add that m time-units from the present, it is a brute fact that time has come to an end. This grounding explanation presupposes that if time does not go on indefinitely, there is a last moment of time. This is not an entirely innocent assumption; however, note that it is compatible with time's being discrete, dense or continuous.

Consider then future contingents of the form $F\varphi$. Suppose first that time will go on for at least n units of time for any number n. Then the truth of $\neg F\varphi$ can be reduced to the truth of the general statement $\forall n F_n\neg\varphi$. To explain how the truth of generalisations may be grounded is a vexed issue; and we do not have anything new to contribute to the solution of the problem. However, once the obligation has successfully been discharged to explain how, for any given choice of n, the truth of $F_n\neg\varphi$ may sometimes be grounded, whatever difficulties remain in explaining how $\forall n F_n\neg\varphi$ might ever be grounded would seem to have nothing specifically to do with the grounding-problem posed by the truth of future contingents.[5] Suppose then that *it is not the case that* time will go on for at least n units of time for any number n. Assuming again that, if this is so, there is a last moment of time, one can hold either that $\neg F\varphi$ is true because time has already come to an end, where it is a brute fact that time has come to an end, or that, for some $0 < m$, $\neg F\varphi$ is true because (i) m time-units from the present, time has come to an end, where m time-units from the present, it is a brute fact that time has come to an end, and (ii) for all k, with $0 < k \leq m$, $F_k\neg\varphi$. The latter is a restricted generalisation, and our previous remarks about the grounding of unrestricted generalisations of type $\forall n F_n\neg\varphi$ apply *mutatis mutandis*.

The problem of accounting for the truth of generalisations, and for that of certain negative statements such as negative existentials, is a problem for everyone; and to the best of our knowledge, the principle of bivalence has never come under attack for these sorts of reasons, except perhaps from intuitionists who presuppose an

[5] The same applies to the problem of truth-grounding for certain negative statements, such as e.g. negative existentials, that we have conveniently glossed over by simply assuming that, in general, if $F\neg\varphi$ is presently true, sometimes in the future, its past truth will be grounded in whatever then grounds $\neg\varphi$.

epistemic conception of truth-grounds quite alien to the metaphysical discussion in which we are here engaged.

Modulo a solution to these latter problems, which as argued are not specific to the debate about the open future, and contrary to what Broad (1923), Dummett (2004) and others contend, GBT accordingly coheres with the principle of bivalence as applied to future contingents, which also silences the worries voiced by critics like D. C. Williams (1951b). Once the grounding requirement on truth is relaxed in these ways – see (GR) above – it will likewise apply to statements about the past: even if, according to GBT, there exist things entirely located in the past, truths about them were, in the past, grounded in how these things then were; and presentists, who deny that there exist such things, may correspondingly say that statements about the past were, in the past, grounded in what there then was and how it then was. On neither view does the grounding requirement, properly construed, force commitment to Lucretian or Cambridge-like properties, while the principle of bivalence can, to this extent at least,[6] be retained.

7.3 Indeterminism and the Open Future

As we have argued, proponents of GBT are entitled to assume classical logic, and in particular a bivalent semantics, even for future contingents. For the same reasons, there is no pressure to abandon the conception of time as being linear in favour of a conception of time as forward-branching. In the light of the appropriately relaxed grounding requirement on tensed truths, proponents of GBT can also draw a systematic distinction between being determinate in truth-value and being predetermined to be true or false: a statement like 'One day hence, some rain will fall' may not be predetermined insofar as its truth is not rendered inevitable by how things located in the present or past of now are or were in all their natural respects; but for all that, it may nonetheless be determinate in truth-value.

Proponents of GBT will therefore refuse to construe the open future in terms of any present lack of truth-value of a certain subset of statements about the future. Instead, they might construe the open future just as the phenomenon that certain statements about the future are neither predetermined to be true nor predetermined to be false – precisely those statements about the future that qualify as future contingents. Given how predetermination was being characterised above, this is also a way in which eternalists might characterise the open future. Even if always, everything always exists, and even if always, everything has its properties permanently – either because the properties in question are relations to times or because the things in question are of point-sized duration – one may nonetheless still allow that how things located in the present or past of now are, in relation to times identical to or earlier than now, does not predetermine how things located in the future of now are in relation to times later than now.

[6] Bivalence may, of course, be said to fail for other reasons, e.g. vagueness.

This may invite the following challenge: if the best that a proponent of GBT can do in order to account for the openness of the future is to claim something that can likewise be claimed by eternalists – i.e. that what will happen is not predetermined in the aforementioned sense – then they have not succeeded in capturing any interesting sense of openness. Even if we consider this too strong a claim, we take up the challenge and show that there is a more radical sense in which the future may be said to be open which is available to GBT, but not to permanentism, and hence, since eternalism implies permanentism, not to eternalism either. This sense is still perfectly compatible with the unrestricted principle of bivalence. And so, even if it should after all be true that if the future is open in any interesting sense, eternalism is incompatible with the future's being open, this thought alone does nothing to motivate the contention that future contingents are neither true nor false. Before we explain what this stronger sense is, however, let us first set out the conception of the open future that both proponents of GBT and permanentists might agree on.

The idea that the future is open is closely related to the rejection of determinism, which latter we may informally gloss as the doctrine that always how the world is, and has been, nomologically determines how the world is going to be. This thought is aptly captured by the following principle (here and below we use special variables t, t', etc. for times, which could be dispensed with, using regular variables and restricting quantification with the help of our predicate T for times):

(DET) Always, $\forall t \mathrm{G} \forall t'$ ($t < t' \rightarrow$ the way the world is up to t nomologically determines the way the world is up to t')

Indeterminism is correspondingly understood as the negation of (DET), i.e. as the claim that

(IND) Sometimes, $\exists t \mathrm{F} \exists t'$ ($t < t'$ & The way the world is up to t does not nomologically determine the way the world is up to t')[7]

(IND) is oftentimes deemed insufficient for the open future, precisely because, as we shall see shortly, on some natural interpretations of it, (IND) is acceptable to both static and dynamic permanentists.[8] Even if this is so, however, there are other interpretations of (IND) that are not acceptable to permanentists.

[7] The formulation of (IND) allows that certain ways the world could have turned out to be were more probable than others given the way the world was up to an earlier time, and so that the laws of nature are probabilistic; and surely, any non-zero probability at least requires nomological possibility.

[8] Barnes and Cameron (2009) have even argued that a thesis like (IND) is not necessary for the open future, contrary to what we suggest here. Their reason for this claim is that it may, in some sense of 'metaphysically indeterminate', be metaphysically indeterminate what the world is like up to now, so that even fully deterministic laws will only take us from the present indeterminate world state to a later indeterminate world state: if there is any indeterminacy in the present world state, this indeterminacy simply 'may bleed over' into the subsequent world state. However, the authors' reasoning seems flawed because the sense in which the present state of the world might be indeterminate – say,

To see this, note first that both (DET) and (IND) are still pretty vague and allow for different precisifications of the phrase 'the way the world is up to t'. We take the latter to be a nominalisation of a statement of the form 'At t, ω^{t}' where 'ω^{t}' is a description of what *actually* is or used to be the case at t, where this description is in some pertinent sense complete. Different precisifications specify different such senses.

It should anyway be clear, though, that lest (DET) be trivialised, 'ω^{t}' must not contain clauses equivalent to clauses of the forms 'Fφ' or '$F_n\varphi$' (for some or all n), which are irreducibly future-tensed. Similarly, 'ω^{t}' must not contain clauses ultimately reducible to clauses of the forms '$\neg F\varphi$' or '$\neg F_n\varphi$' (for some or all n). To see this, consider a case in which a present configuration of particles and force fields determines that in n units of time, a certain event e with a certain determinable property Φ will occur, while no conjunction of present physical facts determines which determinate of Φ e will then instantiate. Suppose that in n units of time, e occurs and instantiates determinate property Φ_1. It would be cheating to try to restore coherence of this example with (DET) by conceiving 'ω^{now}' to entail '$\neg F_n\Phi_i(e)$', for each determinate of Φ, Φ_i, that is distinct from Φ_1.

Bearing these bans in mind, we can now consider one precisification of 'ω^{t}' according to which it yields a complete description of all the entities actually located at t, or at any time earlier than t, in terms of the natural – non-Lucretian and non-Cambridge-like – properties and relations among them that they actually instantiate at t or at any earlier time. To disambiguate, let 'ω^{t}_{loc}' be the relevant description. Since, as far as (DET) is concerned, absences may well be determining factors, we will here presume that 'ω^{t}_{loc}' may contain a suitable number of clauses of the form 'and that's everything located at the present time', either unembedded or embedded in the context of operators of the forms 'P', 'H' or 'At t'', with $t' \prec t$. The corresponding determinist claim then is

$$(\mathrm{DET_{loc}}) \quad \text{Always}, \forall t \mathrm{G} \forall t' \, (t \prec t' \rightarrow \Box(\text{At } t, \omega^{t}_{loc} \rightarrow \text{At } t', \omega^{t'}_{loc}))$$

where '$\Box$' encodes nomological necessity. Both static and dynamic permanentism and GBT are compatible with the truth of ($\mathrm{DET_{loc}}$).[9] Let us therefore now turn to the question of whether either of these two views can be made to cohere with the corresponding form of indeterminism, i.e.

the sense in which it is indeterminate which, if any, cell has survived fission – is not the sense in which the future is said to be indeterminate by being open: by the authors' own lights, the future is open in a sense in which the present and past are not (see Rosenkranz 2013: 69, for discussion).

[9] McTaggart (1927: §337) rightly observed that this would commit proponents of GBT to the truth of statements about the future – something at odds with what Broad takes himself to be committed to (Broad 1923: 73). As we have argued above, however, Broad (1923) is mistaken when he contends that proponents of GBT must treat statements about the future as being neither true nor false: even future contingents can be regarded as bivalent, and some as true, quite consistently with GBT (cf. also Broad 1937: 206).

(IND$_{\text{loc}}$) Sometimes, $\exists t \mathrm{F} \exists t' (t \prec t' \,\&\, \Diamond(\text{At } t, \omega^t_{loc} \,\&\, \neg \text{ At } t', \omega^{t'}_{loc}))$

For simplicity's sake, assume that *now* $\prec t$, and that while actually both 'Now, ω^{now}_{loc}' and 'At t, ω^t_{loc}' hold, '$\Diamond$(Now, ω^{now}_{loc} & ¬At t, ω^t_{loc})' also holds. In application to this particular case, we can distinguish at least three grades of (IND$_{\text{loc}}$) depending on the counterfactual scenarios that can consistently be claimed to bear witness to the possibility claim, *viz.*

(I) Now, ω^{now}_{loc}, but at t, some of the things actually located at t are located at t and have properties distinct from the ones they actually have at t

(II) Now, ω^{now}_{loc}, but while t, and times later than t, sometimes exists, always, for all times t', such that either $t = t'$ or $t \prec t'$, no entity distinct from t' is ever located at t'

(III) Now, ω^{now}_{loc}, but neither t nor any time later than t ever exists

It would seem that both static and dynamic permanentism and GBT can allow for all three grades of (IND$_{\text{loc}}$). Note, however, that on any eternalist ontology exclusively of spatiotemporally individuated events that also have all their non-spatiotemporal properties essentially, (I) is after all no option. Such an ontology would still allow for (III) to hold, though, and also for (II) to hold provided that such an ontology likewise includes instants of time.

According to another precisification of 'ω^{t}', it yields a complete description of all the entities actually *existing* at t, or at any time earlier than t, in terms of the natural – non-Lucretian and non-Cambridge-like – properties and relations among them that they actually instantiate at t or at any earlier time. To disambiguate, let 'ω^t_{ex}' be the relevant description. Since, as far as (DET) is concerned, absences may well be determining factors, we will here presume that 'ω^t_{ex}' may contain a suitable number of clauses of the form 'and that's everything existing at the present time', either unembedded or embedded in the context of operators of the forms 'P', 'H' or 'At t''', with $t' \prec t$.

On this precisification, the corresponding determinist and indeterminist claims accordingly are:

(DET$_{\text{ex}}$) Always, $\forall t \mathrm{G} \forall t' (t \prec t' \rightarrow \Box(\text{At } t, \omega^t_{ex} \rightarrow \text{At } t', \omega^{t'}_{ex}))$
(IND$_{\text{ex}}$) Sometimes, $\exists t \mathrm{F} \exists t' (t \prec t' \,\&\, \Diamond(\text{At } t, \omega^t_{ex} \,\&\, \neg \text{ At } t', \omega^{t'}_{ex}))$

These readings of determinism and indeterminism may admittedly be less familiar; however, they still fit the moulds of (IND) and (DET) and involve admissible precisifications of the phrase 'the way the world is up to t'.

Assume that *now* $\prec t$, and that while actually both 'Now, ω^{now}_{ex}' and 'At t, ω^t_{ex}' hold, '$\Diamond$(Now, ω^{now}_{ex} & ¬At t, ω^t_{ex})' also holds. Again, in application to this particular case, we can distinguish three grades of (IND$_{\text{ex}}$) depending on whether the counterfactual scenarios described by (I′) to (III′) below can consistently be claimed to bear witness to the possibility claim '◊(Now, $\omega^{now}{}_{ex}$ & ¬At t, $\omega^{t}{}_{ex}$)':

(I′) Now, $\omega^{now}{}_{ex}$, but at t, some of the things actually located at t are located at t and have properties distinct from the ones they actually have at t

(II′) Now, $\omega^{now}{}_{ex}$, but while t, and times later than t, sometimes exists, always, for all times t', such that either $t = t'$ or $t \prec t'$, no entity distinct from t' is ever located at t'

(III′) Now, $\omega^{now}{}_{ex}$, but neither t nor any time later than t ever exists

As in the case of (I), on any eternalist ontology exclusively of spatiotemporally individuated events that also have all their non-spatiotemporal properties essentially, (I′) is unavailable. For permanentists, and hence eternalists, (II′) is no option, unless they presume that all things other than times that are actually located at t, or at a time later than t, can all be located at times earlier than t – which would seriously constrain their ontology. Since according to permanentists, now, t and all times later than t exist, (III′) is no option for them, irrespective of what further assumptions they might be willing to make about the existents that they are, *qua* permanentists, committed to.

By contrast, GBT is unproblematically consistent with (I′) to (III′). So there is at least one interpretation of (IND) – (IND_{ex}) of grade (III′) – which coheres with GBT but is clearly not open to permanentists, and another – (IND_{ex}) of grade (II′) – which coheres with GBT but is very unlikely to be compatible with permanentism.

Accordingly, if the task was to come up with a conception of the open future available on GBT but unavailable on any permanentist views, the conception that allows for the possibility of either type of scenario will fit the bill. And yet, this conception of the open future is perfectly compatible with the bivalence of all future-tensed statements, including future contingents.

Presentists, too, can avail themselves of these conceptions of the openness of the future, without being forced to reject, or systematically restrict, bivalence. If they do so, however, they will be forced to say that the past is open in the very same sense, in which case they can no longer account for the asymmetry between the future as open and the past as fixed which underlies the intuitions that fuel attempts to come up with such conceptions in the first place (see, however, Markosian 1995). GBT, of course, faces no such problem, because on GBT, what existed in the past still exists. This makes GBT much better positioned to invoke (IND_{ex}) of grade (II′) or (III′) in the attempt to give an account of the open future.

Chapter 8
Classical Theories of Time, and Relativity

Abstract In this chapter we explicate the challenge posed to classical theories of time by relativistic physics, and show that two recent attempts to reconcile such theories with Special and General Relativity founder. We conclude that a systematic revision of the classical theories is called for. In Sect. 8.1 we argue that the challenge is best conceived as threatening the intelligibility of the postulate, common to all classical theories, that there is an absolute and total temporal order. We show that C. Bourne's appeal to primitive tenses is insufficient to avert the challenge. In Sect. 8.2 we scrutinize D. Zimmerman's recent attempt to construe the postulated temporal order as being imposed by the contents of spacetime rather than its structure. We argue that this attempt fails to answer the challenge, and conclude in Sect. 8.3 that metaphysicians should move on and devise successor theories that no longer postulate such an order.

In this chapter and the next, we will be concerned with the relation between classical theories of time, of which GBT is just one, and modern physics after Einstein. It is a widely perceived view that this relation is, on the face of it at least, highly problematical, even if there may be disagreement about the exact nature of the tension, if any, and even if there may be disagreement about whether there is any serious tension at all that, for the naturalistically minded philosopher, would put classical theories of time in jeopardy.

In this chapter we discuss what might be termed the *conservative strategy*: the strategy, namely, to reconcile classical metaphysics of time with the results of relativistic physics. We identify what we take to be the main challenge posed by relativistic physics, and then examine, in its light, two recent attempts to implement the conservative strategy. The upshot of our discussion will be that these attempts to establish a peaceful coexistence fail to take the sting out of the challenge, and do so in a way that makes it doubtful what, in general, the prospects for the conservative strategy are. We close by advocating an alternative approach which we call the *revisionary strategy*. The next chapter will then see us apply this revisionary strategy to GBT and other classical theories of time.

The plan for the present chapter is as follows. In Sect. 8.1, we argue that the challenge that relativistic physics holds in store for classical theories of time is best conceived as one of intelligibility. Classical theories of time posit an absolute and

F. Correia, S. Rosenkranz, *Nothing To Come*, Synthese Library 395,
https://doi.org/10.1007/978-3-319-78704-6_8

total temporal order, describable by employment of absolute temporal notions. These notions, insofar as they are applicable to the physical world as intended, should have application conditions that in principle allow for characterisation in physically acceptable terms. Yet, given relativistic physics – whether based on Special Relativity (SR) or General Relativity (GR) – it is highly doubtful whether there are any application conditions of this kind, at least whether there are any such conditions whose obtaining is not already ruled out by SR and GR. We then critically discuss Craig Bourne's proposal to define absolute simultaneity in tensed terms and, in light of this challenge, find it wanting.

In Sect. 8.2, we turn to Dean Zimmerman's more recent attempt to address the challenge of intelligibility head-on. Zimmerman distinguishes between what is intrinsic to the structure of spacetime, on the one hand, and the contents that occupy spacetime, on the other, and then argues that while the former may be just the way relativistic physics takes it to be, an absolute and total temporal order may nonetheless be determined by the latter. This implementation of the conservative strategy promises to deliver physically specifiable application conditions for absolute temporal notions in terms of the contingent contents of spacetime, and to make room for their fulfilment while leaving the relativistic conception of the structure of spacetime intact. However, as we try to substantiate in what follows, even this more ambitious project fails, because even if the contingent contents of spacetime allow for a foliation of spacetime into spacelike hypersurfaces that is, in some sense, special, it remains as yet open what the physically specifiable conditions are that such a foliation has to meet in order to count as determining an absolute and total temporal order.

We conclude that, in the light of these failures, the prospects for the conservative strategy look somewhat dim. In Sect. 8.3, we accordingly propose a radical change of tack: instead of trying to reconcile classical theories of time with relativistic physics, we might be better off discerning the kernel of truth that would survive, should relativistic physics win the day and those theories had to be stripped off their commitment to an absolute and total temporal order. This revisionary strategy will require, for its successful implementation, the conception of a language that is not time- but spacetime-sensitive. With hindsight, it can be assumed that relativistic counterparts to classical temporaryism will treat expressions of the form '*m* exists' as being, in the relevant sense, spacetime-sensitive and to vary in truth-value across spacetime. This invites the question of whether, given the temporaryists' original motivations, for them, discerning such theories, and showing them to be consistent with relativistic conceptions of spacetime, is worth the trouble. We make a first stab at answering this question.

In the next chapter, we then give a more rigorous characterisation of the spacetime-sensitive language and implement the revisionary strategy, in order to win through to relativistic versions of GBT and the other classical theories of time.

8.1 Relativistic Physics and an Absolute and Total Temporal Order

Relativistic physics would appear to pose a threat to the classical metaphysics of time. We say 'would appear' because there is an ongoing debate about whether this threat is merely apparent, about how serious the challenge is if there really is one, and about whether the tension between relativistic physics and classical metaphysics of time, if such there be, is eased by reflections on the epistemic status of relativistic physics itself. However, it is not hard to see that, on assumption that relativistic physics gives a correct account of the physical world, there is at least a *prima facie* case for thinking that it is in conflict with traditional theories of time.

As far as relativistic physics is concerned, any total temporal order is only ever relative to a foliation of spacetime into spacelike hypersurfaces. Moreover, in the Minkowskian spacetime of SR, there is a plethora of equally admissible such foliations with nothing recognised by the theory itself to break the tie. Thus, in SR, each inertial frame of reference delivers one such foliation, while all physical laws are invariant across these different frames. According to GR, nothing in the theory itself dictates that spacetime be foliable into spacelike hypersurfaces at all; and even if there is such a foliation in the actual world, there is no guarantee that it has no equally viable alternatives, and no guarantee either that it in any way corresponds to the absolute and total temporal order that prerelativistic physics had taken for granted (for a discussion of the latter predicament, see Bourne 2006).

However, traditional metaphysical theories such as presentism and GBT presuppose that there *is* such an absolute and total temporal order: if whatever exists in space and time is contemporaneous with, or precedes, now, then insofar as it is an absolute fact of the matter what exists, it must likewise be an absolute fact of the matter what is contemporaneous with, or precedes, now (Putnam 1967; Rietdijk 1966; Prior 1970). But not only temporaryist theories presuppose absolute notions of contemporaneity and precedence. To the extent that the intelligibility of temporal operators like 'Always' presuppose a temporal (and not just spatiotemporal) ordering, static permanentism likewise presupposes that it is an absolute fact of the matter what stands in such B-relations to what.

Accordingly, the *prima facie* tension with relativistic physics might then, in a first go, be expressed as follows: if relativistic physics is true, and if it tells us all there is to be told about spacetime, then classical theories of time, such as temporaryism and the B-theory, rest on a presupposition that cannot be redeemed.

Commentators who harbour sympathies for such theories of time are quick to point out that these are two rather big 'if's (Bourne 2006; Zimmerman 2011). On the one hand, they argue that SR has anyway been superseded by GR, so that tensions with SR need not be too disconcerting; and even if GR should in the end provide no safer haven for classical theories of time, its conflict with Quantum Mechanics is yet to be resolved, while it is currently uncertain whether or not the unified Theory of Everything will eventually rehabilitate the idea of an absolute and total temporal order (Zimmerman 2011).

On the other hand, it has been suggested that even if relativity should survive, so that it remains the case that physics itself has nothing to say that would vindicate assumption of such an absolute and total temporal order, physics may just not tell us all there is to be told about spacetime (Prior 1970; Bourne 2006; Zimmerman 2008: 219–20). Thus, it might be argued that common sense or metaphysics, or a combination of the two, reveals to us that there is an absolute and total temporal order, even if we have no means to know, say, whether 'the *n*th pulsation [of a distant body] and the perception of [its] *n*-1th pulsation are simultaneous', since for this we would, plausibly, need to appeal to the theories of physics (Prior 1970: 248; see also Bourne 2006). Alternatively, it might be argued that while according to relativistic physics there is indeed nothing intrinsic to the structure of spacetime that would determine a privileged foliation, both SR and GR, as theories primarily about this structure, are as yet silent on whether the contents of spacetime rather than its structure do not, after all, single out a privileged foliation that would deliver an absolute and total temporal order (Zimmerman 2011).

As to the first line of response, let us observe that even if it should be conceded that GR has superseded SR – and there is still a minority of physicists contesting this assessment of the situation – in any case, it is doubtful whether GR is any more hospitable to such an order, and if one is in a state of uncertainty as to whether a presupposition of one's metaphysical theory will eventually be redeemed, one cannot just lean back and hope for the best. Instead, one should prepare for the worst and contemplate what one will be able to say, consistent with the central tenets of one's view, should it turn out that the Theory of Everything fails to rehabilitate, in physically respectable terms, an absolute temporal order. This takes us to the second line of response.

We begin by noting that it would be a gross understatement of the situation to suggest 'that all that [relativistic] physics has shown to be true or likely is that in some cases we can never *know*, we can never *physically find out*, whether something is actually happening or merely has happened or will happen', e.g. whether 'the *n*th pulsation [of a distant body] and the perception of [its] *n*-1th pulsation are simultaneous' – 'not just simultaneous from such and such a point of view or in such and such a frame of reference, but simultaneous' (Prior 1970: 248). The threat rather is that there might be no way to express, in the language of physics, the conditions under which two such events are, in this sense, absolutely simultaneous – at least none that stand a chance of being fulfilled. This threat is not just an epistemological one, but ultimately one of physical unintelligibility. As such, it should not be taken lightly. For, after all, if the physical intelligibility of the notion of absolute simultaneity is at stake, it is no help to insist that there might be facts readily statable by means of that notion which are, alas, physically undetectable.

Bourne (2006) takes refuge to the fact that, as Prior (1970) observes, physics itself has no use for the tenses. The intelligibility of ordinary tensed discourse would indeed seem beyond dispute, even for the physicist. Bourne's idea is that while relativistic physics may merely traffic in temporal notions that are inherently subject to relativisation, tensed discourse, with which we all are familiar, employs absolute such notions. Accordingly, or so the thought goes, insofar as the tenses are both familiar and absolute and can be used to articulate relations of absolute simultaneity, no threat of unintelligibility looms. In this spirit, and taking a leaf from McTaggart (1927) and Prior (1967), Bourne (2006) offers the following definition of absolute simultaneity for point-sized events:

(B*) *m* is absolutely simultaneous with $n \equiv_{\text{df}}$ P(*m* occurs & *n* occurs) ∨ (*m* occurs & *n* occurs) ∨ F(*m* occurs & *n* occurs)

However, the hope that this might already do in order to give a physically respectable sense to talk about absolute simultaneity, in a way that still allows the notion to have application, is misguided. If, according to relativistic physics, it is only ever relative to a foliation of spacetime that it makes physical sense to ask whether two events *m* and *n* are contemporaneous, or whether they stand in relations of precedence, then from the standpoint of relativistic physics, it is likewise only ever relative to such a foliation that it makes physical sense to ask whether *m* and *n* both presently occur, or whether sometimes in the past, they both occurred, or whether sometimes in the future, they will both occur. In other words, according to relativistic physics, tensed statements of the types exemplified by.

m occurs & *n* occurs
P(*m* occurs & *n* occurs)
F(*m* occurs & *n* occurs)

where in each case, 'occurs' is in the present tense, are to be relativised to foliations of spacetime, inasmuch as those statements are that belong to the types exemplified by:

$m \approx n$
$m < n$

What is more, depending on the foliation of spacetime assumed, provided that there is more than one, the truth-values of statements of either type will vary; and as long as there is nothing that might privilege one such foliation over any other, from the standpoint of relativistic physics, as yet no clear physical sense can be attached to the idea that such statements are ever true *simpliciter*.

Commenting on Putnam's contention that absolute simultaneity must 'be definable in a "tenseless" way in terms of the fundamental notions of physics' (Putnam 1967: 241), Bourne writes:

> [R]equiring that the definition [of absolute simultaneity] be given in tenseless terms is an unargued for assumption that I see no compelling reason to adopt, especially from the point of view of the tense theories; and the same goes for formulations in terms of 'fundamental notions of physics'. Rather, the important issue is whether tense as traditionally conceived is *compatible* with [SR], whether or not it can be formulated in terms of it. What matters is that a convincing definition of [absolute] simultaneity can be given. (Bourne 2006: 174)

He goes on to say that, equally, if one were to issue 'a demand for a *naturalistic* basis for [absolute] simultaneity', one would be 'guilty of begging the question' against dynamic theories of time (Bourne 2006: 174).

However, note first that even if, as argued, the relativistic physicist inevitably reads the tensed statements that figure in the *definiens* of (B*) as being subject to relativisation to foliations of spacetime, this does not in any way suggest that she gives them a *tenseless* reading: even if relativized to a particular foliation, the truth-value of any of the disjuncts in the *definiens* of (B*) – e.g. 'P(*m* occurs & *n* occurs)' – will still vary across different hypersurfaces of that foliation. Secondly, while it may be right to reject the demand that the notion of absolute simultaneity be *definable* in terms of physical theory, its supposed applicability to the physical universe ought at least to be taken to imply that there are physically specifiable conditions of its application; and as long as it is unclear what those physically specifiable application conditions are, the question of whether (B*) is compatible with SR remains open.

Recognising that it is in any case not sufficient to simply lay down (B*), Bourne writes:

> The burden is [...] on me to show how we can understand [the defined] notion of absolute simultaneity, especially given that we can never know which events are absolutely simultaneous with which [...]. It is not good enough to say boldly that we just *do* have some sort of understanding of the notion of absolute simultaneity; that, after all, was what Einstein was dissatisfied with. The question is, then, what does it take to understand it? There are two components: how to understand *simultaneity*; and how to understand *absoluteness*. My solution accepts the challenge that our understanding of simultaneity must be tied to our definition of simultaneity [...]. I suggest this: first, we do understand what it is for ourselves to be *absolutely present* and for present-tense [d] propositions to be absolutely true, for it is not possible for us to be anything but correct about whether we are present, if we are presentists: if we exist, we are present. Second, simultaneity is defined in terms of the conjunction of present-tensed propositions. Thus:
>
> 1. We can understand what it is for something to be simultaneous because we understand the notion of conjunction, the definition of which is entirely exhausted by the truth-table for '&'.
> 2. We understand the content of the present-tense[d] propositions involved in such conjunctions by grasping what they represent. For example, I grasp the present-tense[d] proposition that I am sitting because I know what I am and what it is to be sitting.
> 3. I grasp the notion of the present-tense because it is that of which I have immediate acquaintance.
>
> Thus, I can understand the notion of absolute simultaneity, since there is nothing more to understanding this than understanding the notion of the conjunction of absolutely true present-tensed propositions. (Bourne 2006: 175–76).

Thus, Bourne concedes that in order to properly understand the *definiens* in (B*), in the sense in which it is intended, one must have a grasp of 'absolutely true present-tensed propositions'. Yet, he fails to acknowledge that in order for (B*) to do its job, the past- and future-tenses must likewise be taken in an absolute sense – 'P' and 'F' are mysteriously absent from his list; and he does not acknowledge either that it will not do simply to assume that 'we do understand what it is for ourselves to be *absolutely present*', in order to win through to an understanding of what, in general, it is for present-tensed propositions to be absolutely true – in particular, if these propositions concern the occurrence of spatially remote events.

Accordingly, what has yet to be shown is that the application conditions of the *definiens* of (B) can be given a physically respectable sense that begs no questions, where, unlike (B*), (B) renders the appeal to absolute tenses explicit:

(B) m is absolutely simultaneous with $n \equiv_{\text{df}} \mathscr{T}\text{P}(m \text{ occurs} \ \& \ n \text{ occurs}) \vee \mathscr{T}(m \text{ occurs} \ \& \ n \text{ occurs}) \vee \mathscr{T}\text{F}(m \text{ occurs} \ \& \ n \text{ occurs})$

In the Minkowskian spacetime of SR and in foliable spacetimes of GR, there is a clear and respectable sense in which, at a given point in spacetime, an event may be absolutely present, absolutely past or absolutely future: m is absolutely present/past/future just in case m is present/future/past relative to *all* foliations of spacetime into spacelike hypersurfaces. The present-, past- and future-tenses may then be given a corresponding reading, using operators such as 'Everywhere in the absolute past' and the like. So, proponents of SR or GR may have no trouble understanding the significance of talk about the truth *simpliciter* of tensed statements, as long as they are allowed to give it the following gloss:

(A*) $\mathscr{T}\varphi \rightarrow \forall x(x \text{ is a foliation} \rightarrow \text{Relative to } x, \varphi)$

But then, given the Minkowskian spacetime of SR and spacetimes of GR that are multiply foliable, if (A*) holds, according to (B), only those point-sized events m and n will ever be absolutely simultaneous that are located at the same spacetime-point – which is clearly not what Bourne and other traditional metaphysicians of time intend. But if (A*) does not encode what is intended, how else can it be shown that the absolute tenses Bourne needs meet the test of physical intelligibility?

It would accordingly be a mistake to suppose that tensed language is completely insulated from the effects of relativity, so that one could simply take it for granted that the tenses have an absolute sense and that, since they anyway do not figure in the theories of relativistic physics themselves, their intelligibility alone ensures that they can be used to capture features of reality for metaphysics to speculate about, for which relativistic physics has no implications. Rather, the right to the claim that the tenses can be used in an absolute sense, and as such are both perfectly intelligible and suitable to express an absolute and total temporal order even in the light of modern physics, must be earned.

All this is not to say that whatever makes sense must figure in physical theory or be *analysable* in terms that do. But, to insist, for expressions that purport to describe the *physical* world, it should at least in principle be possible to give a rough characterisation of their application conditions in such terms, where these conditions, thus characterised, may after all obtain. Thus, when Zimmerman (2008: 219), drawing the analogy with tense, rhetorically asks: 'Should we conclude that, since physics does not mention things like dogs, there is no reason to believe in such things – as opposed to mere swarms of particles arranged in various canine shapes?', we may concede that this conclusion is not forced upon us. Yet, the crucial contrast is that, vagueness aside, we have at least a rough idea of which physically specifiable conditions must obtain in order for there to be a *dog* – swarms of particles arranged in certain ways – whereas, given relativity, there is so far no reason to believe that we can come up with a correspondingly physically acceptable specification of the application conditions for tensed expressions construed as absolute, without thereby implying that they remain unfulfilled.

It now transpires that proponents of dynamic permanentism, insofar as they too find (A*) wanting, are likewise under the obligation to avert the threat of physical unintelligibility. For, although in formulating their view, they do not employ any temporal relations, they do use the tenses to articulate tensed truths *simpliciter*. Quite generally, though, to the extent that the meaning of temporal operators calls for a temporal ordering, articulated in terms of B-relations, static permanentists are no better off, unless they can win through to an alternative formulation of their view that makes no use of 'Always' and, unlike the proponents of the classical B-theory, manage to forego any attempt to 'detense' tensed statements by merely finding another argument place for times to fill.

In the attempt to come up with an account of the truth *simpliciter* of tensed statements, proponents of dynamic theories of time, or of the classical B-theory, are more likely to invoke the following characterisation instead:

(A) $\mathcal{T}\varphi \leftrightarrow$ Relative to f, φ

where f is the 'privileged' foliation. Accordingly, the application conditions for 'absolutely simultaneous' can straightforwardly be characterised as follows, sidestepping (B):

(AS) m is absolutely simultaneous with $n \leftrightarrow$ there is a member of f on which both m and n are located

where f is the privileged foliation. Similarly, and exploiting the causal structure attributed to spacetime by SR and GR alike, we can now specify the application conditions for 'absolutely precedes':

(AP) m absolutely precedes $n \leftrightarrow$ there is a member of f, h, on which m is located, and another member of f, h', on which n is located, such that some point on h' is causally accessible from some point on h

where, again, f is the privileged foliation.

Evidently, the question now is what is here meant by 'privileged', and what guarantees that there is a unique privileged foliation in the intended sense of 'privileged'. Proponents of classical theories of time that presuppose an absolute and total temporal order may insist that the meaning of 'privileged' cannot be given save by means of the kind of absolute temporal notions at work on the left-hand sides of (AS) and (AP). Thus, for example, temporaryists may explicate 'the privileged foliation' by saying that it denotes the foliation along which the 'wave of becoming' propagates, in the direction imposed by the causal structure of spacetime (cf. Zimmerman 2011). Since the challenge is *not* to provide an *analysis* of absolute notions in the terms that figure in relativist physics, this is all good and well. But these philosophers cannot thereby escape the challenge to at least devise a physically respectable characterisation of the application conditions of those absolute temporal notions, or the notion of a privileged foliation, for that matter, even if such application conditions do not purport to capture their meaning.

8.2 On Privileging a Foliation: Contents *vs* Structure

In relativistic physics, there is nothing intrinsic to the structure of spacetime that would privilege any foliation of spacetime into spacelike hypersurfaces. Thus, as indicated, in SR, there is a plethora of equally admissible foliations into hyperplanes of simultaneity, corresponding to different inertial frames of reference of objects in relative motion, with nothing in the theory itself to privilege one foliation over any other. The structural constraints identified by GR, by contrast, do not even imply that spacetime be foliable into spacelike hypersurfaces at all; *a fortiori* they do not single out any such foliation as in some sense privileged either.

But may it not be that while the *structural* constraints SR and GR impose on spacetime do not yet call for a privileged foliation, its contingent *contents* ultimately do? SR is silent on what the spacetime manifold contains; and while the structure of the spacetimes of GR does depend on such contents – e.g. the existence of a planet at a certain region of spacetime affects the structure of that spacetime – GR is as yet silent on whether the contingent contents of the actual spacetime do not after all make one of its foliations into spacelike hypersurfaces (if any) in some pertinent sense special. Accordingly, SR and GR might allow for the manifold to be filled with existents in such a way that a privileged foliation emerges after all – even if such existents could have been absent, or been distributed in different ways, so that another foliation, or in fact no foliation at all, would have been singled out. As long as the contents of spacetime that effect such a privileged foliation admit of a physically acceptable description, as they plausibly do, the threat of physical unintelligibility would seem averted.

This is the line of thought that Zimmerman (2011) sets himself the task to explore. Focusing on SR, he writes:

> There are possible distributions of matter in a space-time with Minkowskian metrical properties that are consistent with SR, despite the fact that they effectively 'privilege a foliation'. Here is what should be an uncontroversial example: suppose there were, spread evenly throughout the cosmos, a kind of particle every member of which is moving inertially and at rest relative to every other. This family of fellow-travelers would select an inertial frame; and there would be exactly one foliation of the Minkowskian manifold in which every slice is orthogonal to the path of every one of the special particles. [...] The family of particles is, by hypothesis, very special; and the frame they pick out is, for that reason, also special.

> Should we say that any physical theory that posited such particles would be inconsistent with SR? If, according to the theory, the particles just *happen* to be traveling together in this way, then surely not. So long as the choice of their inertial frame is a contingent matter determined by initial conditions, it should not be attributed to space-time itself, even if they *must* travel on parallel paths. The particles choose a set of parallel inertial paths, and make these paths and the accompanying foliation special, but there need be nothing intrinsically special about the paths in virtue of which the particles *must* take them, rather than those of some other inertial frame. (Zimmerman 2011: 209-10)

This toy example illustrates how the contingent contents of the Minkowskian manifold may make a specific foliation into spacelike hypersurfaces *special*, without relying on there being any feature intrinsic to the structure of the manifold itself that would single out that foliation.

In the end, Zimmerman is after a different such foliation, *viz.* the one that presentism is committed to: a foliation into spacelike hypersurfaces such that, for any spacetime-point, whatever exists in space and time at that point lies on the hypersurface that is determined by this foliation to be the one on which that very point lies. For Zimmerman, presentism is the only real contender among the various dynamic theories of time: he thinks to be able, not only to set aside the Moving Spotlight Theory, but also to reject GBT for the kinds of reasons that we critically reviewed in Chap. 6 and found wanting. But for present purposes it is just as well to follow Zimmerman here and explore how the presentist might fare when she invokes the idea of a foliation as being privileged by the contents of the manifold. The lessons to be learnt will in equal measure affect GBT. Like Zimmerman, we will begin by assuming the Minkowskian spacetime of SR and say something about spacetimes obeying the constraints of GR later.

Already in the description of the toy example, we can see a problem emerging that will, or so we shall argue, ultimately deal a deathblow to this attempt at giving physically respectable sense to an absolute and total temporal order. For, while there is no doubt that the family of particles that the example invokes would make the foliation into spacelike hypersurfaces which are orthogonal to their paths in some sense *special*, this may not be sufficient to make that foliation *privileged* in the sense of 'privileged' that the friend of absolute simultaneity targets when she lays down (A), (AS) and (AP).

To fix ideas, let us assume that there is indeed a foliation of Minkowskian spacetime into spacelike hypersurfaces of the kind the presentist needs; and for ease of exposition, and despite Zimmerman's declared neutrality on the issue, let us assume that these hypersurfaces are indeed hyper*planes*. By hypothesis, the hyperplanes do not intersect, and so each spacetime-point lies on one and only one such hyperplane. Accordingly, given this, there is assumed to be a foliation f that is special in the following sense: for any given spacetime-point s, at s, everything in spacetime is located somewhere on the spacelike hypersurface which is determined by f to be the one on which s itself lies.

It follows that for any spacetime-points s and s', if at s, s' exists, and at s', m exists, then at s, m likewise exists – accordingly, existing at a spacetime-point is *transitive*. It likewise follows that, at any spacetime-point s, no region of spacetime exists that is, relative to the time-axis determined by f, earlier or later than the hypersurface on which *s* lies. The latter is of course nothing physicists are likely to accept. Yet, the characterisation of *f* only uses notions that, from the standpoint of relativistic physics, are perfectly respectable; and positing a foliation that satisfies this characterisation would seem perfectly consistent with the structure of Minkowskian spacetime, irrespective of whether physicists are ready to endorse that there *is* such a foliation.

So, *if f* is identified with the privileged foliation in the sense of (A), (AS) and (AP), the 'wave of becoming' can be said to propagate along the time-axis determined by *f*, and so time passes along this axis, as slices of the manifold successively come into and go out of existence. But what does this suggestion amount to? What is the physical net value of the privilege that this identification bestows? Of course, presentists, just as proponents of any of the other theories of time that presuppose an absolute and total temporal order, will insist that the suggested identification has a clear sense; and we cannot, without begging the question, assume that, here, they are subject to an illusion. However, even if we concede, for the time being, that there *are* coherent such absolute notions of simultaneity and becoming, the price of this concession is that it now becomes an open question whether *f* – never mind its being in some pertinent ontological sense special, courtesy of the contingent contents of the manifold – is the privileged foliation in the sense of 'privileged' at work in (A), (AS) and (AP). The contingent contents of the Minkowskian manifold alone do not answer *this* question. Zimmerman writes:

> Positing a wave of becoming inevitably privileges a single foliation; nevertheless, if the laws determining its location do not themselves appeal to non-Minkowskian space-time structure, the privileging does not require that the foliation be special in-and-of-itself – in advance of the contingent conditions that choose one foliation to be the lucky winner. A wave of becoming that obeys this law requires no more help from the manifold than the family of particles envisaged earlier: particles that inevitably move inertially and at rest relative to one another, but that could have been introduced into space-time in any frame. (Zimmerman 2011: 217)

However, the contingent conditions at most select which foliation, if any, is such that, for any given spacetime-point *s*, at *s*, everything in spacetime is located somewhere on the spacelike hypersurface that is determined by that foliation to be the one on which *s* itself lies. To be the 'lucky winner' of *this* contest is not *eo ipso* to be the foliation that 'positing a wave of becoming [...] privileges'. To determine the 'location' of the foliation that enjoys the latter privilege, the laws to which Zimmerman alludes must employ the notion of a privileged foliation, or the absolute temporal notions in terms of which that notion might be taken to be defined: the laws must, precisely, state that any foliation is privileged which is such that for any given spacetime-point *s*, at *s*, everything in spacetime is located somewhere on the spacelike hypersurface that is determined by that foliation to be the one on which *s* itself lies – and this claim cannot be taken to yield a stipulative definition of 'privileged', lest it be vacuous and incapable of determining anything at all. As such, it is as yet unclear whether these laws do not, after all, presuppose something about the intrinsic structure of spacetime at odds with relativistic physics. In any case, however, the worry has not yet been allayed that the absolute temporal notions,

Fig. 8.1 Two foliations of spacetime

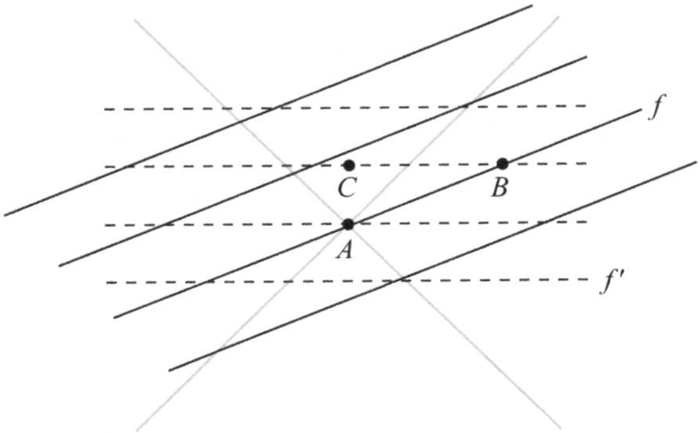

including the notion of an objectively privileged foliation of the kind the 'wave of becoming' requires, are, physically speaking, unintelligible. Thus, to the extent that the relevant laws employ those very notions, the question has not gone away what the physical net value of the privilege is that these laws bestow.

To appreciate the problem, see Fig. 8.1 and consider the foliation f' (dotted lines) that cuts across foliation f (plain lines), which latter we continue to assume satisfies the aforementioned characterisation.

Now assume what the friend of an absolute and total temporal order has no means to discard as unintelligible, *viz.* that f' rather than f is privileged so that, according to temporaryism, the 'wave of becoming' propagates along the time-axis of f' and, in line with (AS) and (AP), absolute simultaneity/precedence coincides with simultaneity/precedence relative to f'.

What would such a scenario imply? For instance, at any spacetime-point, there would then be things space-like separated from that point that are absolutely earlier or absolutely later than that point. Similarly, at a given spacetime-point – say *A* – there would then exist a thing – say *B* – that is space-like separated from, but absolutely later than *A*, while nothing space-like separated from, but absolutely contemporaneous with *B* existed. For instance, at *A*, *C* would not exist. Moreover, there would then be, at a given spacetime-point – say again *A* – some thing – say again *B* – that is space-like separated from *A* such that *A* absolutely precedes *B* by *n* units of time (i.e. *n* units of f'-time), for some number *n*, while *n* units of time absolutely later than *A*, there would be no spacetime-point causally reachable from *A* at which *B* existed – in fact, *n* units of time absolutely later than *A*, *B* would *only* exist at the spacetime-point at which it is located. In this way, what exists would not only vary across f'-time but also along the hyperplanes of absolute simultaneity that, *ex hypothesi*, f' determines, and so across space.

One might attempt to invoke principles, statable in neutral terms, in order to rule out a scenario such as this. For example, one might hope that appeal to the aforementioned principle of transitivity will already do:

> For any spacetime-points *s* and *s'*, if at *s*, *s'* exists, and at *s'*, *m* exists, then at *s*, *m* likewise exists.

But then, the envisaged scenario does not violate this principle. Thus, while, as noted, at *A*, (the point occupied by) *B* would exist, but at *A*, *C* would not exist, at (the point occupied by) *B*, *C* does not exist either. In general, transitivity is obeyed, since whatever exists at a spacetime-point *s* also exists at any point at which *s* exists.

Alternatively, one might try to appeal to causal principles such as the principle according to which, for any spacetime-point *s*, at *s*, there used to be, in the past lightcone of *s*, a point *s′* infinitesimally close to *s*, so that whatever happens at *s* may be taken to be, in part or fully, causally determined by what happened at *s′* (cf. Zimmerman 2011: 193–94). But again, the envisaged scenario is consistent with such a principle. To illustrate this, just imagine the distance between the *f*-hypersurfaces on which *A* and *C* respectively lie to be infinitesimally small; then whatever happens at *C* may, fully or in part, be causally determined by what happened at *A*, even if it was *f′*, and not *f*, that determined the absolute and total temporal order.

How else might the presentist rule out that *f′* rather than *f* is the privileged foliation, and so the foliation along whose time-axis the 'wave of becoming' propagates? Consider the following constraint:

(C) ∀*x*(*x* is a spacetime-point → At *x*, ∀*y*∀*n*(Relative to the privileged foliation, *x* precedes *y* by *n* time-units → Relative to the privileged foliation, *n* time-units in the future, everywhere, ∀*z*(*z* is a spacetime point → At *z*, *y* exists)))

To the extent that *f* exists, and so matter is distributed in the way *f* requires, only *f* satisfies this constraint. By contrast, as we have seen, the principle of transitivity, according to which what exists at a given spacetime-point *s* also exists at any spacetime-point at which *s* exists, is underwritten by both *f* and *f′* in our example.

Similarly, no matter whether *f* is privileged or *f′* is privileged, it may be said to be a matter of principle that, for every spacetime-point *s*, what happens at *s* be caused by what happens at a point from which, according to SR and GR alike, *s* can be reached. So (C) looks *prima facie* more promising.

However, unlike the principle of transitivity or principles of causality of the kind mentioned, (C) employs the very notions whose physical intelligibility is in question, in the sense that it remains utterly unclear, from the standpoint of physics, what their application conditions are. Thus, if we replaced 'Relative to the privileged foliation' by 'Relative to *f*' throughout, where *f* is picked out by the characterisation of it that we have given, then (C) would become vacuous and so incapable of imposing any constraint at all. Zimmerman writes:

> What is inconsistent with merely Minkowskian intrinsic structure is to explain some fact about the contents of space-time as being due to the special nature of one foliation, and then not be able to appeal to any deeper laws that fail to mention that foliation. If the laws of a theory merely pick out the relevant frame of reference in terms of contingent material contents, and the contents merely *happen* to pick out that frame; then it is the material contents that are doing the work. But if a theory's most basic laws (whether they govern physical or metaphysical features of the manifold) must invoke one inertial frame of reference or folia-

> tion 'by name', as it were; then there is something special about the frame or foliation itself, quite apart from the manifold's content. The law is an indication that the manifold includes built-in 'rails', directing things in a certain way; some structure that is part of space-time itself is doing the work. (Zimmerman 2011: 213)

Given the alternative Zimmerman here describes, as a genuine constraint, (C) would thus appear to assume '"rails" built into the manifold, telling events which slices they should occupy' – if only against the backdrop of the thesis that there *is* a privileged foliation in whose terms the absolute and total temporal order can be conceived. At least, the need to invoke (C), or a similarly suitable constraint that rules out f', shows that it is simply not the case that 'the "rails" [are] laid down by the way matter is distributed' (Zimmerman 2011: 215). It is therefore difficult to see how (C) can be made to square with the contention that it is the 'material contents that are doing the work' *rather than* 'some structure that is part of space-time itself'.

(C) may not 'invoke one inertial frame of reference or foliation "by name"' (whatever exactly this would come to); but this does not alter the fact that it overtly appeals to the privileged foliation, provided that such a privileged foliation exists – which latter proviso is, by the presentist's own lights, satisfied. Zimmerman (2011) is primarily concerned to show that the assumption of such a privileged foliation can be rendered consistent with relativistic physics; but as long as it is unclear what physical basis, if any, *this* assumption has – rather than merely the assumption that there is a special foliation such as f – its consistency with SR cannot be ascertained. If it could be presupposed that the very notion of a privileged foliation need have no physical net value at all – as opposed to a metaphysical one, say – then the consistency of this assumption with SR could, of course, be earned on the cheap, since, in general, it is uncontroversial that a theory T is consistent with another theory T' as long as T's subject matter is disjoint from the subject matter with which T' deals, and T and T' are themselves individually consistent. Thus, for example, to say that objects meeting a certain physical condition are blessed by the gods is uncontroversially consistent with physics, as long as it can be taken for granted that being blessed by the gods is not itself a physical property or a property partially grounded in physical properties. However, such a take on the notion of a privileged foliation lacks any plausibility: it is meant to apply to the very spacetime that physics studies.

We therefore conclude that, given SR, the challenge to show that assumption of an absolute and total temporal order has some basis in physical fact, cannot be answered in the way Zimmerman proposes, *viz.* by appeal to the contingent contents of spacetime. This assessment of the situation does not change in any relevant way, once we turn to GR. For, even if it transpired that our universe, governed by the constraints of GR, did not allow for any other foliation save the special one that *would fit the presentist's bill, if it were privileged* – so that we cannot use the same ploy as before in order to illustrate the problem – that problem would nonetheless still be there: it still needs to be explained what basis in physical fact the very privilege has that presentist philosophers bestow on that foliation. Evidently, the same would apply, *mutatis mutandis*, to proponents of GBT.

8.3 Preparing for the Worst: The Revisionary Strategy

For all that has been argued here, classical theories of time might ultimately prove consistent with relativistic physics – or at least they might, for all we presently know, be reinstated by a future Theory of Everything that supersedes relativistic physics. However, given what has been argued here, there is no reason either for optimism that classical theories and relativistic physics can happily coexist; and given our current epistemic situation, it might just as well turn out that no future Theory of Everything will dethrone relativistic conceptions of spacetime. To prepare for the worst, in such a context, is to prepare for a case in which classical theories of time do *not* cohere with relativistic physics, while the latter still survives, and so for a scenario in which the following pessimistic assumption holds:

(Pess) In assuming an absolute and total temporal order, classical theories of time presuppose non-relativistic physics, while there is no rehabilitation of non-relativistic physics on the horizon

If (Pess) holds, the conservative strategy is doomed. However, not all is lost in such a case, if it can nonetheless be shown that each classical theory of time can be purged of whatever it is that makes it committed to prerelativistic physics – and this in such a way that a theory remains that is recognisable as its relativistic counterpart. Thus, if we can find a q, inconsistent with relativistic physics, such that for any classical theory T, there is a relativistic theory T_R consistent both with prerelativistic physics and with relativistic physics such that if q holds, T is true if T_R is true, then *modulo* (Pess), we can conceive of T as decomposable into two independent parts, q and T_R, such that even if q fails, T_R may nonetheless survive relativity. The search for such a q and such theories T_R is what the *revisionary strategy* recommends.

To have any chance of success, implementation of this strategy is subject to certain constraints. First, the relevant claim q must allow for articulation in terms acceptable from the point of view of relativistic physics, lest it remain unclear in what sense the latter is inconsistent with the former. Secondly, the relevant relativistic theories T_R must themselves allow for such an articulation, lest it remain unclear in what sense they might be said to cohere with relativistic conceptions of spacetime, whereas their prerelativistic counterparts do not. Thirdly, in order for us to be in a position to assess whether if q holds, T indeed is true if T_R is true, the language in which q and T_R are couched must be suitably related to the tensed language of T.

To tackle these issues, our lead idea is to devise a spacetime-sensitive language, definable in relativistic terms, that allows us to express T_R, and to choose q in such a way that upon the supposition that q holds, the time-sensitivity of the language of T correlates in appropriate ways with the spacetime-sensitivity of the language of T_R.

This is still very unspecific; but we will render this idea more precise in the next chapter (as well as in Appendix 2 and Appendix 3). That there should be a spacetime-sensitive language definable in relativistic terms at all should not be surprising. The causal structure of the manifold, posited by both SR and GR, determines that what is causally reachable from one point in spacetime differs from what is causally reachable from another; and so there is no principled objection against the thought that there may be relativistically acceptable expressions that are true at one spacetime-point but not at another.

It is of course a further, more contentious claim that expressions of the form '*m* exists' are of this spacetime-sensitive kind; and with hindsight, it is clear that certain of the T_R will subscribe to this controversial claim, *viz.* those T_R whose prerelativistic counterparts are versions of temporaryism. This invites the general question to what extent, if any, such T_R will be faithful to the ideas originally motivating their prerelativistic counterparts, and so be worth the rescue. In the remainder of this chapter, we will try to make a first stab at answering this question.

It is not as if variation in what exists across different spacetime-points is altogether alien to prerelativistic temporaryism. In prerelativistic physics, a spacetime-point can be conceived of as a pair of a space-point and a time; and according to prerelativistic temporaryism, there *is* variation in what exists across spacetime-points, thus conceived, *viz.* those that differ in their time. Temporaryists do, however, refuse to admit any such variation in what exists across spacetime-points which share their time and hence are absolutely simultaneous in the sense prerelativistic physics assumes. The assumption that all spacetime-points can thus be divided into equivalence classes of absolutely simultaneous spacetime-points is what underlies their contention that what exists varies with time but not with space. But once this assumption founders, as (Pess) suggests, and no objective and absolute time-coordinates can be assumed to be shared by distinct spacetime-points, the contention has no longer any clear sense; and we lose any principled way to distinguish between those spacetime-points across which there is such variation and those across which there is none.

But then, those who originally harboured sympathies for temporaryism are faced with a choice: either they see failure of the assumption of absolute simultaneity as a reason to deny, *tout court*, that there is any variation in what exists across different spacetime-points; or else they see failure of that assumption as a reason to adopt the view that, generally, what exists varies from spacetime-point to spacetime-point.

The first option would force them to say that, at a point where a certain event occurs, the event's causal effects likewise exist, despite the fact that it is only at distinct, time-like separated points that those effects unfold. But if there ever was any attraction to temporaryism in the first place, this idea should seem unpalatable. If one ever thought, in a prerelativistic setting, that there is nothing yet to come, one should continue to think, in a relativistic setting, that there is nothing wholly in the causal future of here-now. Similarly, if one ever was persuaded, in a prerelativistic setting, that there is nothing purely past, one should continue to think, in a relativistic setting, that there is nothing wholly in the causal past of here-now.

What the second option would force friends of temporaryism to accept ultimately depends on the shape the relevant relativistic theory T_R takes; but we can at least say this much: they would have to accept that even at non-causally separated points (i.e. points that are causally insulated from one another) – if any two distinct such points exist – different things exist. Combined with the relativistic insight that what is (not) causally reachable anyway varies across distinct spacetime-points, this is not equally obviously in tension with the original attraction to temporaryism – it only would be in such a tension, if it could be assumed that there was after all a sense in which non-causally separated points could be said to be absolutely simultaneous. But *ex hypothesi* the latter assumption fails; and once this is taken to heart, the motives for temporaryism may at best be said to fall silent on the matter. In fact, however, if at all spacetime-points, there are several distinct non-causally separated points, then given the relativistic insight, the thought that at all spacetime-points, the causal future is empty should drive one towards accepting that at different non-causally separated points, different things exist.

On balance, therefore, it is fair to say that those who had sympathies for prerelativistic temporaryism, once forced to accept (Pess), should be more inclined to choose the second option than to open the floodgates and accept that what exists at any given point likewise exists at any other. It is in the light of this diagnosis that we will explore the prospects of the revisionary strategy in the next and final chapter.

Chapter 9
Spatiotemporaryism

Abstract In this chapter we devise a spacetime logic and argue that temporaryism must give way to spatiotemporaryism, which latter construes variation in what exists as variation across spacetime. In Sect. 9.1 we argue that much of the rationale for thinking, in a prerelativistic setting, that what exists varies across time, should survive the finding that there is no absolute and total temporal order and rationalise the corresponding thought that what exists varies across spacetime. In Sect. 9.2 we introduce a spacetime-sensitive language and a spacetime logic with operators and relations defined over the fourfold causal structure of spacetime. In Sect. 9.3 and Sect 9.4 we use these tools to articulate and compare competing spatiotemporaryist ontologies and contrast them with spatiopermanentism according to which what exists does not vary across spacetime. In Sect. 9.5 we show which relativistic views can naturally be taken to correspond to which classical theories of time.

In this chapter, we explore the implementation of the revisionary strategy. This strategy steers clear of any pretence that relativistic physics may after all cohere with the assumption of a privileged foliation of spacetime into spacelike hypersurfaces. Instead, it seeks to devise alternative versions of classical theories of time that, unlike their prerelativistic predecessors, are consistent with failure of that assumption.

To implement this strategy, we must first come to terms with the fact that, failing such a privileged foliation, there is no way to identify an absolute and total temporal order. In particular, we can no longer assume that there exist times – in any absolute sense in which truths about existence are truths *simpliciter* – and that each such time divides all others exhaustively into those in the past of it and those in the future of it. Instead, we must do justice to the thought that, as Minkowski famously remarked in 1908, 'space by itself, and time by itself, are doomed to fade away into mere shadows, and only a kind of union of the two will preserve an independent reality' (Minkowski 1952: 75). In other words, we must make spacetime-points our basic points of evaluation and appeal to no objective structure of spacetime other than the fourfold division, determined by each such point, between that point itself, those points in the causal past of it, those in the causal future of it, and those in the elsewhere region of it.

F. Correia, S. Rosenkranz, *Nothing To Come*, Synthese Library 395,
https://doi.org/10.1007/978-3-319-78704-6_9

Variation in what exists, if any, must accordingly be understood as variation across spacetime rather than time. Hence, temporaryism must give way to what we might call *spatiotemporaryism*, and in order to fully articulate this generic view, as well as its more specific varieties, we must ultimately use a language that is spacetime- rather than time-sensitive – at least, we are bound to do so if we are intent on identifying metaphysical theories that might, with reason, be conceived as relativistic counterparts to GBT, presentism and dynamic permanentism. This, in turn, makes it mandatory to devise such a language in the first place, equipped with spatiotemporal operators whose behaviour is correspondingly regulated by what we might call a *spacetime logic*. Once such a language and such a logic are to hand, we can go on to formulate and evaluate competing spatiotemporaryist ontologies, and contrast them with the *spatiopermanentist* view according to which, everywhere in spacetime, what exists also exists everywhere else in spacetime.

But before embarking on this enterprise, we must address the more general worry that a variation in the facts across spacetime simply is unintelligible, or at least borders on the absurd – to the extent, namely, that this would have been our verdict in the prerelativistic setting, had it initially been suggested instead that the facts vary across space.

Our plan for this chapter accordingly is as follows. In Sect. 9.1, we address the general worry just rehearsed and attempt to defuse it by further reflecting on the causal structure of the manifold at each spacetime-point. In Sect. 9.2, we give an informal characterisation of a spacetime-point-sensitive language, including a set of spatiotemporal operators, which proves adequate for the description, at each spacetime-point, of the causally structured spacetime manifold. We go on to devise a spatiotemporal logic that mimics, as closely as possible, the tense logic which we introduced in Chaps. 1, 2 and 3. In Sect. 9.3, we then use these resources to distinguish between spatiotemporaryism and spatiopermanentism, and set out those spatiotemporaryist views that are the most natural candidates for succeeding the temporaryist ontologies described in Chaps. 4 and 5. As it turns out, there are two relativistic successor theories for each of presentism and GBT. In Sect. 9.4, we compare the costs and benefits of these four competing ontologies, single out one of each pair of candidates as superior to its respective rival, and conclude with a brief assessment of one of the most striking consequences that the two superior spatiotemporaryist theories share, *viz.* that what exists at a spacetime-point existing at s may not exist at s itself. Appendix 2 offers a semantic characterisation of each of the systems identified. Finally, in Sect. 9.5, we identify two bridge principles that, once conjoined with the relativistic views, yield the corresponding prerelativistic views, given a suitable translation function from the prerelativistic language into the relativistic language. Taken together, these two principles amount to the prerelativistic assumption of an absolute and total temporal order as determining the sole dimension along which facts may be said to vary. The effect of conjoining relativistic theories with these two bridge principles confirms our earlier suspicion that only one of each pair of successor theories, corresponding to GBT and presentism respectively, is a plausible candidate for serving as the relativistic counterpart to the prerelativistic view it succeeds, being most faithful to the latter's rationale. Appendix 3 renders these considerations precise.

9.1 Causal Structure and Factual Variation across Spacetime

In Special Relativity (SR) and General Relativity (GR) alike, for each spacetime-point *s*, there is a fourfold, exhaustive division of the manifold into *s* itself and three disjoint regions: the causal past of *s*, the causal future of *s*, and the elsewhere region of *s*. The causal past of *s* is the region of spacetime-points, if any, happenings at which can causally affect what happens at *s*. The causal future of *s* is the region of spacetime-points, if any, happenings at which can be causally affected by what happens at *s*. The elsewhere region of *s*, at last, is the region of spacetime-points, if any, non-causally separated from *s*, happenings at which can neither causally affect, nor causally be affected by, what happens at *s*. The simplified Minkowski diagram in Fig. 9.1 illustrates this causal structure for the special case of SR.[1]

Here, the causal future includes the surface of the light cone represented by the ∨-shaped region, minus *s*; and *mutatis mutandis* for the causal past and the surface of the ∧-shaped region, minus *s*.

This causal structure, though relative to a given spacetime-point, does *not* depend on any further parameter, such as, for instance, a foliation of spacetime or a reference frame. Accordingly, *if* the facts change from spacetime-point to spacetime-point – so that sentences of the form 'Here-now, φ' may be used to express truths *simpliciter* – then, likewise, there will be truths *simpliciter* expressible by sentences of the forms 'Somewhere in the causal past, φ', 'Somewhere in the causal future, φ', and 'Somewhere in the elsewhere region, φ'. Indeed, *if* the facts can change from spacetime-point to spacetime-point, then such changes can only properly be articulated by using sentences of such forms. The pressing question is, however, whether

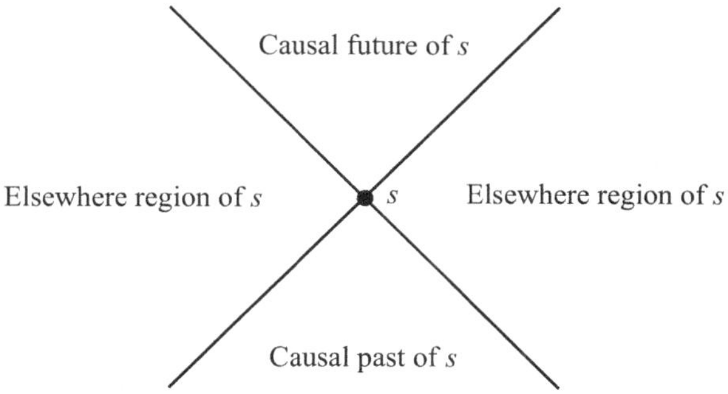

Fig. 9.1 A spacetime point and its surroundings

[1] The causal structure of Lorentzian manifolds is a topic of its own in relativistic physics. A classic is Chap. 6 of Hawking and Ellis (1973), and a recent survey is provided by Minguzzi and Sánchez (2008).

the thought that the antecedent condition be satisfied is so much as coherent. To fix ideas, and with hindsight, let us focus on facts about what exists.

To begin with note that, for any spacetime-point *s*, what happens at spacetime-points, if any, in the causal past of *s* can still be said to *precede* what happens at *s* – in the very sense of 'precede' at work when we say that causes precede their effects. Similarly, what happens at spacetime-points, if any, in the causal future of *s* can still be said to *succeed* what happens at *s* – in the very sense of 'succeed' at work when we say that effects succeed their causes. That there be variation in what exists across spacetime-points ordered in this way does not become an unintelligible or absurd suggestion, just because what is ordered are spacetime-points rather than time-instants. What *is* puzzling, by contrast, is the finding that when we utter 'now' we refer, if to anything at all, to *here-now*, and that we are accordingly faced with the choice of either denying that there are any spacetime-points besides here-now that causally neither precede nor succeed it, or else accepting that the order in terms of causal precedence and succession is not total and admitting a region of spacetime-points non-causally separated from *here-now*, including here-now, across which there likewise is variation in what exists.

To think, in a relativistic setting, that what exists is relative to spacetime-points – including those, if any, populating the elsewhere region of here-now – is arguably no more counterintuitive than to think, in a prerelativistic setting, that what exists is relative to time-instants. For, arguably, any felt *surplus* of counterintuitiveness merely derives from the assumption that one is in a relativistic setting, and hence from SR and GR themselves. Neither SR nor GR score high on the scale of intuitiveness – which is a good reason not to give too much importance to this scale in evaluating physically informed, metaphysical proposals.

To the extent that one had any temporaryist leanings at all, before one learnt anything about relativistic physics, one thought that what exists may depend on times, but not on places. But this is so because one then thought that there was *absolute simultaneity* between what goes on at distinct places, so that a change of place would not imply a change in what exists, as long as such a spatial change would only take one to a place such that anything occurring there is absolutely simultaneous with what occurs at the departure point. Since in learning about relativistic physics, one learns that there is no such absolute simultaneity, one must now ask oneself afresh whether any objection remains to the suggestion that what exists is relative to spacetime-points. We submit that, for a temporaryist at least, coming to terms with relativistic physics still makes spatiotemporaryism a natural enough idea (Stein 1968; cf. also Stein 1970 and 1991 for further discussion).

For instance, prerelativistically, as a proponent of GBT one might have given a causal interpretation of one's view, thinking that what now exists is exhausted by what now happens and by whatever conspired to make or let it happen, or conspired to make or let happen what so conspired. In other words, one might have thought that what exists is located somewhere in a network of causal chains culminating in what happens now. Variation in what exists at locations upstream and downstream of such causal chains was already contemplated by one's proposal. It then comes as a surprise that 'now' refers, if to anything, to *here-now*, and that what one had excluded from one's ontology were not just things located beyond the culmination point. Once one realises that there is a region neither upstream nor downstream of those causal chains, and that one's criterion of existence implies that there is also variation in what exists across that region, why should this finding automatically mandate that one abandon that criterion? Is the criterion rendered unintelligible or absurd by this finding?

Alternatively, one might have thought that what now exists is exhausted by what happens now and by whatever has the potential to conspire with what happens now to make or let happen what has not yet occurred. In other words, one might have thought that what exists is located somewhere in a network of causal chains that are going to culminate beyond the here-now in what has not yet happened. Variation in what exists at locations upstream or downstream of such causal chains was already contemplated by one's proposal. It then comes as a surprise that 'now' refers, if to anything, to *here-now*, and that one had included in one's ontology things with the potential to causally affect what is going to happen much later, yet unlike what happens now, without causally affecting anything that is going to happen before – and accordingly had admitted things located in causal chains not including anything that happens now. Once one realises that there is a region neither upstream nor downstream of causal chains including what happens now, and that one's criterion of existence implies that there is also variation in what exists across that region, why should this finding automatically mandate that one abandon that criterion? Is the criterion rendered unintelligible or absurd by this finding?

We submit that the answers to these questions should be negative, because such charges are either based on considerations that give pride of place to everyday intuitions – considerations that should have already been taken to be undermined by science itself, and should anyway not be considered the best currency for evaluating metaphysical proposals – or else because these charges trade on the unwarranted claim that we lack the conceptual resources to properly articulate the spatiotemporaryist positions they seek to discredit. Here, we shall say no more on everyday intuitions. Instead, we will proceed to provide a framework which allows us to properly articulate, in relativistically acceptable ways, the very spatiotemporaryist proposals under attack, and to trace out their respective implications. Providing such a framework should silence any remaining worries that spatiotemporaryism is unintelligible.

9.2 Spatiotemporal Logic

There is admittedly no unique phenomenon in natural language, akin to tense, that would allow straightforward recapture by means of spatiotemporal operators, i.e. operators that shift the circumstances of evaluation to those prevailing at certain points in spacetime rather than others. However, we are familiar with ordinary statements to the effect that, say, here and now something is going on that was not going on over there before; and indeed, we are familiar with the idea that any context of utterance not only determines a time, but also a place – a feature of the context of utterance to which indexicals like 'here' are sensitive. Accordingly, once we buy the Minkowskian lesson that we ultimately cannot treat the conjunctive phrase 'here and now' as a specification of two independent parameter values, but must instead understand it along the lines of the hyphenated 'here-now' denoting a point in relativistic spacetime, we should not find it too hard to conceive of a language whose sentences φ are such that the equivalence '$\varphi \leftrightarrow$ Here-now, φ' is true on each occasion of its use. Just like 'here' and 'now', 'here-now' is an indexical whose value is always determined by the context of utterance. But we can also think up a nonrigid modifier corresponding to this indexical, e.g. 'Locally-presently', and readily conceive of the sentences of the language in question, φ, as being everywhere in spacetime equivalent to 'Locally-presently, φ'. If this were so, we could introduce operators that shift the spacetime-point of evaluation across spacetime, e.g. 'Somewhere in the causal future', 'Everywhere in the causal past', *etc.* – just as, in the prerelativistic setting, we could introduce temporal operators like 'Sometimes in the future' and 'Always in the past'.

In what follows we shall assume the availability of such a spacetime-point-sensitive language, which we model as closely as possible on the tensed language used in previous chapters. However, while in the prerelativistic setting, we had to deal, at each time, with a threefold division of the elements of the temporal order, we must now that we are in a relativistic setting, countenance at each spacetime-point a fourfold division of the spacetime manifold. This means that in order to reach all regions of the manifold, we need more operators than we used to. In particular, we must invoke operators of the form 'Somewhere in the elsewhere region' and 'Everywhere in the elsewhere region', to which no temporal operators correspond.

Let us write

'$\blacktriangle\varphi$'	for 'Everywhere in the causal past, φ'
'$\blacktriangledown\varphi$'	for 'Everywhere in the causal future, φ'
'$\blacktriangleleft\varphi$'	for 'Everywhere in the elsewhere region, φ'
'$\vartriangle\varphi$'	for 'Somewhere in the causal past, φ'
'$\triangledown\varphi$'	for 'Somewhere in the causal future, φ'
'$\triangleleft\varphi$'	for 'Somewhere in the elsewhere region, φ'
'$\bullet\varphi$'	for 'Everywhere in spacetime, φ'
'$\circ\varphi$'	for 'Somewhere in spacetime, φ'

We take '$\blacktriangle$', '$\blacktriangledown$' and '$\blacktriangleleft$' as primitive and define the other spacetime operators in their terms as follows. First we define '$\vartriangle$', '$\triangledown$' and '$\triangleleft$':

$$\vartriangle\varphi \equiv_{df} \neg\blacktriangle\neg\varphi$$
$$\triangledown\varphi \equiv_{df} \neg\blacktriangledown\neg\varphi$$
$$\triangleleft\varphi \equiv_{df} \neg\blacktriangleleft\neg\varphi$$

Thus, to hold somewhere in the causal past just is not to fail to hold everywhere in the causal past; to hold somewhere in the causal future just is not to fail to hold everywhere in the causal future; and to hold somewhere in the elsewhere region just is not to fail to hold everywhere in the elsewhere region. Next we define '●' and '○':

$$●\varphi \equiv_{df} ▲\varphi \,\&\, \varphi \,\&\, ◀\varphi \,\&\, ▼\varphi$$
$$○\varphi \equiv_{df} △\varphi \vee \varphi \vee ◁\varphi \vee ▽\varphi$$

In other words, being everywhere the case just is being the case everywhere in the causal past, here-now, everywhere in the elsewhere region and everywhere in the causal future, while being somewhere the case just is being the case either somewhere in the causal past, or here-now, or somewhere in the elsewhere region, or somewhere in the causal future.

Below, we will give informal glosses on the formal principles we introduce, using labels that wherever possible recognisably correspond to those we used for tense-logical principles. This means that sometimes the numbering does not match the chronological order in which these principles are being introduced. To avoid unnecessary clutter, we will throughout omit the qualification 'causal'.

The following axioms and rules are supposed to govern the aforementioned spacetime operators:

($A1_R$) $\varphi \rightarrow ▲▽\varphi$
($A2_R$) $\varphi \rightarrow ▼△\varphi$

Whatever is here-now the case is both everywhere in the past somewhere in the future the case, and everywhere in the future somewhere in the past the case. These two axioms mimic the tense-logical axioms (A1), i.e. '$\varphi \rightarrow HF\varphi$', and (A2), i.e. '$\varphi \rightarrow GP\varphi$'.

($A3_R$) $▲(\varphi \rightarrow \psi) \rightarrow (▲\varphi \rightarrow ▲\psi)$
($A4_R$) $▼(\varphi \rightarrow \psi) \rightarrow (▼\varphi \rightarrow ▼\psi)$
(AR1) $◀(\varphi \rightarrow \psi) \rightarrow (◀\varphi \rightarrow ◀\psi)$

If everywhere in the past φ implies ψ, then if φ is everywhere in the past the case, so is ψ; if everywhere in the future φ implies ψ, then if φ is everywhere in the future the case, so is ψ; and if everywhere in the elsewhere region φ implies ψ, then if φ is everywhere in the elsewhere region the case, so is ψ. ($A3_R$) and ($A4_R$) mimic the tense-logical axioms (A3), i.e. '$H(\varphi \rightarrow \psi) \rightarrow (H\varphi \rightarrow H\psi)$', and (A4), i.e. '$G(\varphi \rightarrow \psi) \rightarrow (G\varphi \rightarrow G\psi)$'. The additional axiom (AR1) follows the same pattern.

($A5_R$) $▽△\varphi \rightarrow (△\varphi \vee \varphi \vee ◁\varphi \vee ▽\varphi)$
($A6_R$) $△▽\varphi \rightarrow (△\varphi \vee \varphi \vee ◁\varphi \vee ▽\varphi)$
(AR2) $◁◁\varphi \rightarrow (△\varphi \vee \varphi \vee ◁\varphi \vee ▽\varphi)$

Whatever holds somewhere in the future somewhere in the past, also holds somewhere; whatever holds somewhere in the past somewhere in the future, also holds somewhere; and whatever holds elsewhere elsewhere, also holds somewhere. The first two of these axioms mimic the tense-logical axioms (A5), i.e. '$F P \varphi \rightarrow (P \varphi \vee \varphi \vee F \varphi)$', and (A6), i.e. '$P F \varphi \rightarrow (P \varphi \vee \varphi \vee F \varphi)$'. (AR2) has no tense-logical precursor but a similar rationale.

(AR3) $(\triangleleft \triangledown \varphi \vee \triangledown \triangleleft \varphi) \rightarrow (\triangleleft \varphi \vee \triangledown \varphi)$
(AR4) $(\triangleleft \vartriangle \varphi \vee \vartriangle \triangleleft \varphi) \rightarrow (\triangleleft \varphi \vee \vartriangle \varphi)$

If φ holds either elsewhere somewhere in the future, or somewhere in the future elsewhere, then φ either holds elsewhere or somewhere in the future. Similarly, if φ holds either elsewhere somewhere in the past, or somewhere in the past elsewhere, then φ either holds elsewhere or somewhere in the past.

($A7_R$) $\vartriangle \vartriangle \varphi \rightarrow \vartriangle \varphi$

If somewhere in the past, somewhere in the past, φ holds, then somewhere in the past, φ holds. This axiom mimics the tense-logical axiom (A7), i.e. '$P P \varphi \rightarrow P \varphi$'; it prescribes that, in the Kripke-models to be introduced in Appendix 2, the causal precedence relation between spacetime-points is transitive.

(AR5) $\varphi \rightarrow \blacktriangleleft \triangleleft \varphi$

Whatever is here-now the case is everywhere in the elsewhere region elsewhere the case. (AR5) prescribes that the relation of *being in the elsewhere region of* in those Kripke models is symmetric.

($A14_R$) $\circ \vartriangle \top$, for $\top$ any chosen tautology

Just as the tense-logical axiom (A14), i.e. '$P \top \vee F \top$', ensures that all models have at least two times standing in relations of precedence, this axiom ensures that all models have at least two spacetime-points standing in relations of causal precedence. As such, it averts the collapse of interestingly different metaphysical theories.

($R1_R$) $\varphi / \blacktriangle \varphi$
($R2_R$) $\varphi / \blacktriangledown \varphi$
(RR) $\varphi / \blacktriangleleft \varphi$

For any theorem φ, it is also a theorem that everywhere in the past, φ is the case, that everywhere in the future, φ is the case, and that everywhere in the elsewhere region, φ is the case. The first two rules mimic the tense-logical rules (R1), i.e. '$\varphi / H \varphi$', and (R2), i.e. '$\varphi / G \varphi$'. The spacetime-logical rule (RR) follows the same pattern. These three rules allow us to derive the rule which mimics (RD1), i.e. 'φ / Always, φ':

($RD1_R$) $\varphi / \bullet \varphi$

In other words, if φ is a theorem it is likewise a theorem that everywhere, φ is the case. Similarly, the following theorems can now easily be proved[2]:

- ●(φ → ψ) → (●φ → ●ψ)
- ●φ → φ
- ●φ → ●●φ
- φ→ ●○φ

Accordingly, ● *(everywhere)* and ○ *(somewhere)* behave like □ and ◊ in S5.

In addition to these spatiotemporal operators, we employ spatiotemporal operators of the form '@*m*', abbreviating 'At spacetime-point *m*'. We take 'S' to abbreviate the predicate 'is a spacetime-point' and lay down the following axioms for such spatiotemporal operators which mimic those we postulated for temporal operators of the form 'At *m*':

($A8_R$) @*x*φ → S*x*
($A10_R$) @*x*(φ → ψ) → (@*x*φ → @*x*ψ)
($A11_R$) @*x*¬φ ↔ (S*x* & ¬@*x*φ)
($A12_R$) ●φ → (S*x* → @*x*φ)
($A13_R$) @*x*φ → ●@*x*φ

From ($A12_R$) and ($RD1_R$) we have.

($RD2_R$) φ / S*x* → @*x*φ

and given ($RD2_R$), from ($A8_R$) and ($A10_R$), we also have.

($RD3_R$) φ → ψ / @*x*φ → @*x*ψ

It follows from ($A11_R$) and ($A12_R$), by ($RD3_R$), that what holds at a spacetime-point, somewhere holds:

($T2_R$) @*x*φ → ○φ[3]

Our theory of quantification and logic of identity remain the same as before, with the sole exception that (A16), i.e. 'Sometimes, E!*m*', is replaced by.

($A16_R$) ○E!*m*

[2] (AR1) is used for proving the first item on this list, axioms (AR3), (AR4) and (AR2) are used for proving the third item, and (AR5) for proving the fourth item. The second item on the list already follows from the definition of '●'.

[3] Suppose @*x*φ. Then by ($RD3_R$), @*x*¬¬φ. By ($A11_R$), it follows that S*x* & ¬@*x*¬φ. Using ($A12_R$), we get ○φ.

In other words, somewhere m exists. Of course, unlike before, we now understand claims of existence and non-existence as being spacetime-point-sensitive.

Given this, we can now immediately add the following axiom that mimics the tense-logical (A22), i.e. '$\forall x(\mathrm{T}x \rightarrow \mathrm{At}\ x, \varphi) \rightarrow \varphi$':

(A22$_\mathrm{R}$) $\forall x(\mathrm{S}x \rightarrow @x\varphi) \rightarrow \varphi$ (with x not free in φ)

In other words, if it is the case that, for every spacetime-point, at that point φ holds, then here-now φ also holds. Note that (A22$_\mathrm{R}$) entails that there is at least one spacetime-point: just let φ be some logical falsehood, then '$\exists x\mathrm{S}x$' follows.

We take S to be subject to the following constraints:

(A9$_\mathrm{R}$) $\mathrm{S}x \rightarrow \bullet\mathrm{S}x$
(A23$_\mathrm{R}$) $\mathrm{S}x \rightarrow @x\mathrm{E!}x$

That is to say, *being a spacetime-point* is assumed to be a 'ubiquitous' feature of spacetime-points, just as, in the prerelativistic setting, being a time was assumed to be an eternal feature of times: if something is a spacetime-point, then everywhere it is a spacetime-point. Furthermore, any spacetime-point is assumed to exist at itself.

We need the resources to express and attribute spatiotemporal relations between spacetime-points. We take the relations of spatiotemporal location and causal precedence as primitive, symbolize them by 'LOC' and 'PREC' respectively, and begin by laying down the following axioms for 'LOC':

(A25$_\mathrm{R}$) $x\ \mathrm{LOC}\ y \rightarrow \mathrm{S}y$
(A26$_\mathrm{R}$) $\mathrm{S}x \rightarrow (x\ \mathrm{LOC}\ y \leftrightarrow x = y)$
(A27$_\mathrm{R}$) $x\ \mathrm{LOC}\ y \rightarrow \bullet(x\ \mathrm{LOC}\ y)$

In other words, everything is only ever spatiotemporally located at a spacetime-point; any spacetime-point is located only at itself; and *being spatiotemporally located at* is a 'ubiquitous' feature of things and their locations that does not vary across spacetime: if x is located at y, then everywhere, x is located at y. We allow spacetime-points to be located at themselves. Naturally, and with hindsight, we do not take LOC to be existence-entailing. With LOC in place, we can now define what it is to be in spacetime:

(D20) m is in spacetime $\equiv_{\mathrm{df}} \bigcirc\exists x(m\ \mathrm{LOC}\ x)$

Similarly, we can define what it is to be *a resident of spacetime* as follows:

(D21) $\mathrm{R}_{\mathrm{sp}}m \equiv_{\mathrm{df}} m$ is in spacetime & $\bullet\forall x\bullet(m\ \mathrm{LOC}\ x \rightarrow (\mathrm{E!}x \rightarrow \mathrm{E!}m))$ & $\bullet(\mathrm{E!}m \rightarrow \exists y(m\ \mathrm{LOC}\ y))$

The relation of causal precedence among spacetime-points is governed by the following axioms:

(A28$_R$) x PREC $y \rightarrow (\mathrm{S}x \,\&\, \mathrm{S}y)$
(A29$_R$) x PREC $y \rightarrow \bullet(x$ PREC $y)$
(A30$_R$) $\neg(x$ PREC $x)$
(A31$_R$) $(x$ PREC $y \,\&\, y$ PREC $z) \rightarrow x$ PREC z

Accordingly, causal precedence is an irreflexive but transitive relation between spacetime-points that does not vary across spacetime. Assuming the irreflexivity of causal precedence amounts to excluding closed causal curves, and hence to the requirement that spacetime satisfies the 'causality condition' (see Hawking and Ellis 1973: 190). By contrast, the transitivity of causal precedence is a fact whatever the spacetime. Like spatiotemporal location, causal precedence is not existence-entailing. PREC is exclusively a relation between spacetime-points. However, with both PREC and LOC in place, we can now define a more general notion of causal precedence among things in time which we symbolize by 'PREC$_g$':

(D22) m PREC$_g$ $n \equiv_{df} n$ is in spacetime & $\circ\exists x(m$ LOC $x \,\&\, \bullet\forall y(n$ LOC $y \rightarrow x$ PREC $y))$

We next define the binary relation of *being non-causally separated from*, obtaining between spacetime-points, which we symbolize by 'SEP':

(D23) m SEP $n \equiv_{df} \mathrm{S}m \,\&\, \mathrm{S}n \,\&\, \neg(m = n \vee m$ PREC $n \vee n$ PREC $m)$

Thus, two spacetime-points are non-causally separated just in case they are distinct and none of them causally precedes the other. The relation SEP is irreflexive, but symmetric, and so is not transitive. Like LOC and PREC, SEP is not existence-entailing.

With these notions in place, we can now, lastly, introduce the following four mixed axioms. First consider:

(A33$_R$) $@x\blacktriangle\varphi \leftrightarrow \mathrm{S}x \,\&\, \bullet\forall y(y$ PREC $x \rightarrow @y\varphi)$
(A34$_R$) $@x\blacktriangledown\varphi \leftrightarrow \mathrm{S}x \,\&\, \bullet\forall y(x$ PREC $y \rightarrow @y\varphi)$
(AR6) $@x\blacktriangleleft\varphi \leftrightarrow \mathrm{S}x \,\&\, \bullet\forall y(y$ SEP $x \rightarrow @y\varphi)$

where in each case y is not free in φ and x and y are distinct variables. That is to say, if at a given spacetime-point s, everywhere in the past of that point, φ holds, then everywhere it is the case that φ holds at any spacetime-point causally preceding s, and *vice versa*. If at a given spacetime-point s, everywhere in the future of that point, φ holds, then everywhere it is the case that φ holds at any spacetime-point causally succeeding s, and *vice versa*. If at a given spacetime-point s, everywhere in the elsewhere region of that point, φ holds, then everywhere it is the case that φ holds at any spacetime-point non-causally separated from s, and *vice versa*. The rationale for these principles is similar in kind to the one we already gave for the tense-logical axioms (A33), i.e. 'At x, H$\varphi \leftrightarrow$ Tx & Always, $\forall y(y \prec x \rightarrow$ At y, $\varphi)$', and (A34), i.e. 'At x, G$\varphi \leftrightarrow$ Tx & Always, $\forall y(x \prec y \rightarrow$ At y, $\varphi)$'. Finally consider:

(A24$_R$) $@x\forall y\varphi \leftrightarrow \bullet\forall y@x(E!y \rightarrow \varphi)$

where x and y are distinct variables. In other words, if at a given spacetime-point s, everything is such that φ holds, then everywhere, everything is such that, at s, it either does not exist or φ holds, and *vice versa.* (A24$_R$) corresponds to the tense-logical axiom (A24), i.e. '(At x, $\forall y\varphi) \leftrightarrow$ Always, $\forall y$ At x, $(E!y \rightarrow \varphi)$'.

Equipped with this spacetime-logic, we now turn to the formulation of different ontological views none of which presupposes the existence of a privileged foliation of spacetime into spacelike hypersurfaces.

9.3 Varieties of Spatiotemporaryism

As before in the prerelativistic setting, we can distinguish between two broad classes of views that differ on whether there is factual variation – here: across spacetime rather than time:

(STA$_R$) $\forall p, \bullet(\mathscr{T}p \rightarrow \bullet\mathscr{T}p)$
(DYN$_R$) $\exists p, \circ(\mathscr{T}p \ \& \circ\neg\mathscr{T}p)$

In other words, according to the first type of view, wherever something is true *simpliciter*, and so fully articulates a fact, it also is everywhere else a truth *simpliciter*, while according to the second type of view, somewhere some truth *simpliciter* somewhere else fails to be a truth *simpliciter*.

We continue to make the following two assumptions about truth *simpliciter*:

$\mathscr{T}\varphi \rightarrow \varphi$
$E!m \rightarrow \mathscr{T}E!m$

Given these assumptions, (STA$_R$) can be shown to imply *spatiopermanentism*, i.e. the view according to which everywhere in spacetime, whatever exists also exists everywhere else in spacetime:

(SPER) $\bullet\forall x\bullet E!x$

Thus, assume (STA$_R$). Given the two assumptions above, we can derive '$\bullet\forall x\bullet(E!x \rightarrow \bullet E!x)$'. Thanks to (A16$_R$), we can then derive '$\bullet\forall x\circ\bullet E!x$', and hence, given $\circ\bullet\varphi \rightarrow \bullet\varphi$, '$\bullet\forall x\bullet E!x$'.

Spatiopermanentism contrasts with *spatiotemporaryism*, i.e. the view according to which somewhere in spacetime, something exists that somewhere else in spacetime does not exist:

(STEMP) $\circ\exists x\circ\neg E!x$

There is one variety of spatiotemporaryism that pretty straightforwardly corresponds to presentism as we had defined it in Chap. 5, *viz.* the combination of the following two principles:

($P2_R$) $Sx \rightarrow @x\blacktriangle\neg E!x$
($P3_R$) $Sx \rightarrow @x\blacktriangledown\neg E!x$

In other words, at any spacetime-point *s*, it is the case that everywhere in the causal past of *s*, and everywhere in the causal future of *s*, *s* fails to exist. Accordingly, at *s*, *s* is 'one-off' along the trajectory of any particle passing through *s*.

For want of a better label, let us call this view *spatiopresentism*. That spatiopresentism is a version of spatiotemporaryism can be shown as follows:

Proof: By ($P2_R$) and ($A23_R$), $Sx \rightarrow @x\exists y\blacktriangle\neg E!y$ is a theorem. By ($A22_R$), then, $\exists y\blacktriangle\neg E!y$ is a theorem. Given ($A14_R$), i.e. $\bigcirc\triangle\top$, we can infer $\bigcirc\exists y(\blacktriangle\neg E!y \;\&\; \triangle\top)$. Given that $(\blacktriangle\varphi \;\&\; \triangle\top) \rightarrow \bigcirc\varphi$ is a theorem, we can infer $\bigcirc\exists y\bigcirc\neg E!y$. □

Spatiopresentism, as characterised, is as yet neutral as to whether the elsewhere region of here-now is populated. Accordingly, we may add one or the other of the following principles to the mix, obtaining two different versions of spatiopresentism:

(PO) $Sx \rightarrow @x\blacktriangleleft\neg E!x$
(BO) $Sx \rightarrow @x\blacktriangleleft E!x$

The version of spatiopresentism that results from adding (PO) to ($P2_R$) and ($P3_R$) entails that everywhere in spacetime there is one and only one spacetime-point:

(1) $E!x \;\&\; E!y \;\&\; Sx \;\&\; Sy \rightarrow x = y$

Proof: Assume ($P2_R$), and ($P3_R$) and (PO). Next assume $E!m \;\&\; E!n \;\&\; Sm \;\&\; Sn$. From the second assumption, by ($A22_R$), it follows that there is a spacetime-point, *s*, such that $@sE!m$ and $@sE!n$. From the first and second assumptions and axioms ($A8_R$), ($A11_R$), ($A33_R$), ($A34_R$) and (AR6), both $@sE!m \rightarrow s = m$ and $@sE!n \rightarrow s = n$ follow. By the transitivity of identity, $m = n$ follows. □

Although, to the best of our knowledge, this version of spatiopresentism has never been defended in print, it has been critically discussed by Hinchliff (2000: 579) and Savitt (2000: 567–68). If here-now is the only spacetime-point that there is, it follows that every resident of spacetime is located here-now.

Another version of spatiopresentism results, once we add (BO) to ($P2_R$) and ($P3_R$). This kind of spatiopresentism allows for the elsewhere region of here-now to be populated, i.e. for there to exist a spacetime-point non-causally separated from here-now, and hence equally, for there to exist residents of spacetime that are located in the elsewhere region but are not located here-now.

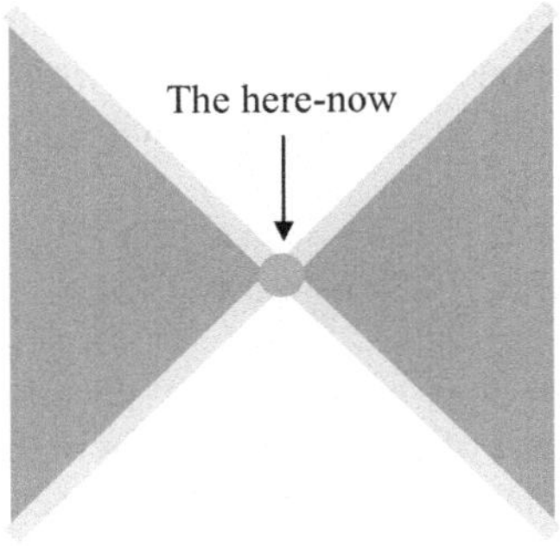

Fig. 9.2 Bow-tie spatiopresentism

This is the kind of view that was put forward, and to a certain extent defended, by Weingard (1972). Focusing on SR, Weingard writes:

> Consider a space-time point P with its associated lightcones. [...] Assuming I-now is real and at P, I should conclude that since I can consider it to be occurring now, any event outside my lightcone can be considered real. But while distant simultaneity is a matter of convention, being real, I take it, cannot be merely a matter of convention. Thus, if an event can be considered real it must be real and so all events outside the lightcone of P (of me-now) are real.
>
> Now in terms of actual physical or experimental facts, it is the class of events that can be considered simultaneous to an event at P, and not the class of events absolutely simultaneous to the events at P, that plays the role in special relativity that the class of events simultaneous to P plays in Newtonian space-time. In each they are the class of events that are not causally connectable with P. And while the class of events simultaneous to an event at P, with respect to some frame of reference, is not a relativistic invariant, the class of events that can be considered simultaneous to events at P is such an invariant. It is just the class of events outside of P's lightcone. Thus, I conclude that the belief that all things that exist now (or are in the present) are real is also true for relativistic space-time, with one qualification. In special relativity the absolute present of an event at P is not the class of events simultaneous with the event with respect to some frame of reference but rather it is the class of events located outside of P's lightcone (plus the events at P). (Weingard 1972: 120-21)

If we understand the regions in dark grey to be those that, according to the view at hand, may be populated, then this version of spatiopresentism can be represented by Fig. 9.2.

By contrast, the spatiopresentist view that results from adding (PO) instead of (BO) would simply have a darkened here-now. We may accordingly call the first version of spatiopresentism *pointy spatiopresentism*, and call the second version *bow-tie spatiopresentism*. We will discuss the relative merits, and relative shortcomings, of these views in Sect. 9.4.

The relativistic counterpart to GBT is naturally conceived to consist in the combination of the following two principles, one of which it shares with spatiopresentism:

($P1_R$) $E!x \rightarrow \blacktriangledown E!x$

($P2_R$) $Sx \rightarrow @x\blacktriangle\neg E!x$

In other words, everywhere everything everywhere in the causal future still exists, while for any spacetime-point *s*, at *s*, everywhere in the causal past of *s*, *s* did not yet exist. Accordingly, at *s*, *s* is 'new' on any particle's trajectory passing through *s*, while it continues to exist on this trajectory even after the latter has passed through *s*. Let us call this kind of view *relativistic GBT*.

That relativistic GBT is a version of spatiotemporaryism has in effect already been established: the earlier proof that spatiopresentism is a version of spatiotemporaryism only made use of ($P2_R$), which it shares with relativistic GBT, and not ($P3_R$), which relativistic GBT rejects.

On relativistic GBT, as on spatiopresentism, there are, at spacetime-point *s*, no spacetime-points in the causal future of *s*, and hence neither any residents of spacetime that, in the generalised sense of causal precedence, are preceded by *s*. Unlike spatiopresentism, however, relativistic GBT contends that, at *s*, there are spacetime-points in the causal past of *s*, and so things in time that, in the generalised sense of precedence, precede *s*.

Nothing has so far been said about the status of the elsewhere region; and we can, as before, distinguish between two versions of relativistic GBT, a pointy and a bow-tie version. Again, the difference between pointy and bow-tie versions depends on which of the following two principles is accepted:

(PO) $\quad \mathrm{S}x \rightarrow @x\blacktriangleleft\neg \mathrm{E!}x$
(BO) $\quad \mathrm{S}x \rightarrow @x\blacktriangleleft \mathrm{E!}x$

With (PO) being added to ($P1_R$) and ($P2_R$), we can derive that, at any given spacetime-point, the elsewhere region is unpopulated:

(2) $\quad \mathrm{S}x \rightarrow @x\neg\exists y(x \text{ SEP } y)$

Proof: Assume Sx and assume for *reductio* $\neg @x\neg\exists y(x \text{ SEP } y)$. Then by ($A11_R$), $@x\exists y(x \text{ SEP } y)$. (AR6) allows us to derive the theorem: $x \text{ SEP } y \ \& \ @y\blacktriangleleft\varphi \rightarrow @x\varphi$. Given (PO), we then derive: $x \text{ SEP } y \ \& \rightarrow @x\neg \mathrm{E!}y$. But then we get $@x\exists y @x\neg \mathrm{E!}y)$. Using ($A24_R$) we can get: $\bigcirc\exists y @x(\mathrm{E!}y \ \& \ @x\neg \mathrm{E!}y)$. Given the logic of @, this yields: $\bigcirc\exists y\bigcirc(\mathrm{E!}y \ \& \ \neg \mathrm{E!}y)$, which cannot be the case. □

The pointy version of relativistic GBT can be represented by Fig. 9.3.

This kind of view was defended by Stein (1968) in response to charges by Rietdijk (1966) and Putnam (1967). Focusing on SR, Stein maintains that the following characterisation of what exists – or in his parlance, of what is real – 'can be taken over word for word from prerelativistic to Einstein-Minkowski space-time; because the relation "being in the past of" is an ordering of space-time, the obvious compatibility requirement – that if *a* has already become for *b*, and *b* for *c*, then so has *a* for *c* – is satisfied':

> For an event – a man considering, for example – at a space-time point *a*, those events, and only those, have already become (real or determinate), which occur at points in the topological closure of the past of *a*. (Stein 1968: 14)

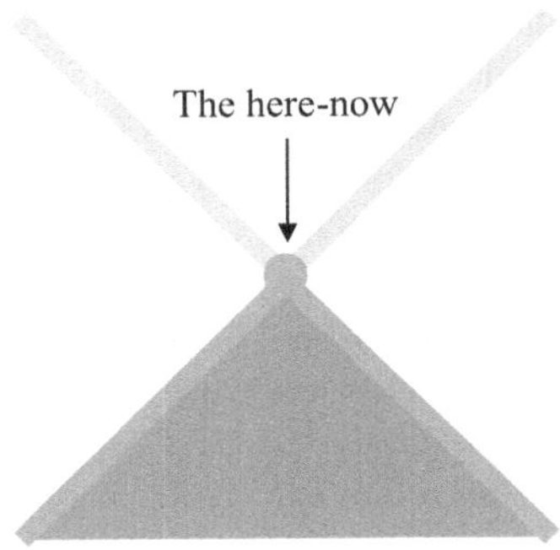

Fig. 9.3 Pointy relativistic GBT

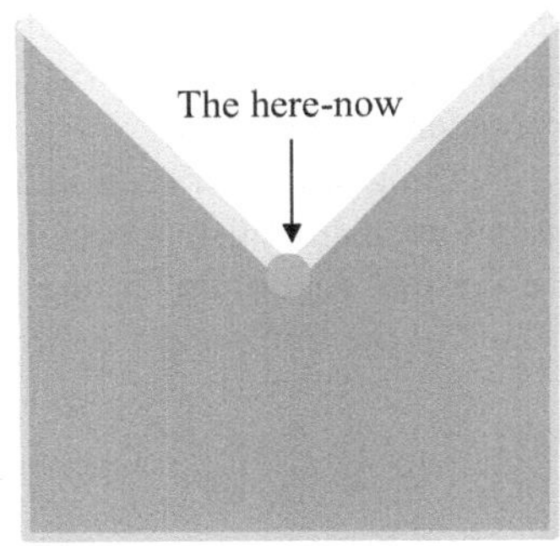

Fig. 9.4 Bow-tie relativistic GBT

He goes on to clarify, in a footnote:

> 'Topological closure' because it is natural to say that something in or on the boundary of the past of *a* has 'by now' become for *a*; note that *a* itself is thereby included (and, for Einstein-Minkowski space-time, besides the past of *a* only *a* itself). (Stein 1968: 14)

By contrast, to the best of our knowledge, the bow-tie version of relativistic GBT has never been defended in print. It results from adding (BO) to ($P1_R$) and ($P2_R$) and can be represented by Fig. 9.4.

It is this view that, as we shall argue in the next section, is the best version of relativistic GBT, just as bow-tie spatiopresentism is superior to its pointy rival. Before we ultimately address these issues, however, let us briefly pause to reflect what *dynamic spatiopermanentism* might look like. As a version of spatiopermanentism, it holds that, everywhere, everything exists everywhere else, and so everywhere takes all regions of the manifold to be populated. However, as a version of (DYN_R), it nonetheless claims that some propositions are somewhere true *simpliciter* without everywhere being true *simpliciter*.

In Chap. 5, we identified what we considered the best version of dynamic permanentism as the kind of view that, for some specific *p*, claims

Ephemerally, p

where ‘Ephemerally, φ’ was defined as ‘Sometimes, (φ & H¬φ & G¬φ)’, and where truths about how far from the present, p is the case, were supposed to be truths *simpliciter*:

$$D_k p \rightarrow \mathscr{T} D_k p$$

Accordingly, as p’s being the case is successively less and less future, and then present, and then more and more past, reality itself changes on this view.

Now, while we could easily redefine ‘Ephemerally, φ’ as ‘○(φ & ▲¬φ & ◀¬φ & ▼¬φ)’, we would need to use a metric tensor in order to express distances from here-now across spacetime. While we will not here be concerned with fleshing out the details, the overall shape of dynamic spatiopermanentism should be clear enough.

9.4 The Intransitivity of Being

We noted before that on the view we called *pointy spatiopresentism*, every resident of spacetime is located here-now. It might be suggested that this commits its proponents to the implausible claim that no resident of spacetime is spatiotemporally extended, i.e. that every resident of spacetime must be point-sized, which would render the view untenable. However, this objection must remain unconvincing. Mereological fusions, if they exist, exist whenever, and wherever, one of their parts exist. Thus, even on the view under attack, residents of spacetime may, for all that, be mereological fusions of spatiotemporal parts, as long as one of their spatiotemporal parts is located here-now. Accordingly, pointy spatiopresentism can after all leave room for the existence of things in spacetime that are spatiotemporally extended.

However, there is another objection, very similar to the one just reviewed, that should carry more conviction (cf. Savitt 2000: 568). In fact, this more powerful objection likewise counts against the pointy version of relativistic GBT, i.e. the kind of view defended by Stein (1968).

Let us, for the sake of argument, engage in the fiction that an embodied consciousness can occupy a single spacetime-point. Here-now, I perceive my different limbs. I furthermore perceive my different limbs to be located at distinct spacetime-points. It is true that, in order for me to perceive my limbs, light signals must reach me, which takes time. So strictly speaking, I only ever perceive that, somewhere in the immediate causal past, my limbs occupy distinct spacetime-points (cf. Russell 1912: Chap. 3). Unlike here-now, my causal past is a region big enough to comprise several spacetime-points and so has the ‘thickness’ that past location of my different limbs requires. However, according to pointy views, *at no point* in my causal past are there sufficiently many points in order for all my different limbs to be located, not even if we think of each of my limbs as a mereological fusion of residents of spacetime, where such fusions are located wherever any of their parts are located. Since light signals may simultaneously reach me from different points in my causal past, this need not yet be a problem for pointy views. However, I have every reason to believe that my limbs have persisted. In order for them to have done so, they would have to here-now exist, and so would some points in the causal future of the points in my causal past where I here-now perceive them to have been located. For, *ex hypothesi*, my limbs are fusions of residents of spacetime, and wherever the latter exist, so do points at which they are located. But my different limbs are not all located here-now; so where are they located, if they here-now still exist? It is no consolation to be told that at least on pointy relativistic GBT, as opposed to pointy spatiopresentism, my limbs still exist, for while persistence implies existence, existence is quite compatible with a failure of having persisted – in the sense of persistence at issue here, *viz.* that of continuing to exist while located.

Note that bow-tie views face no comparable problem, because both on bow-tie spatiopresentism and bow-tie relativistic GBT, whenever I here-now perceive my limbs to have been located at points in the causal past of here-now, there exist, here-now, several spacetime-points in the causal future of those points. We take these considerations to disclose a clear advantage of bow-tie views over their pointy rivals.

However, it has been argued that both varieties of bow-tie spatiotemporaryism run afoul of certain principles that are either claimed to be independently plausible or claimed to encode intuitions that these views are meant to do justice to. Thus, Savitt (2000) argues against views such as Weingard's on the following grounds:

> First, it seems reasonable to require that [event] *E* itself be a member of any set *S* of events representing the present for *E* in [Minkowskian spacetime]. Second, it also seems reasonable to require that, if some set of events *S* represents the present for event *E*, then no events in *S* should be in each other's absolute past or absolute future (that is, it should not be the case that all observers at *E* agree that one of the events is, say, earlier than the other). Let us call this the requirement that any set of events representing the present in [Minkowskian spacetime] must be *achronal*. (Savitt 2000: 567)

Savitt then argues that on a view such as bow-tie spatiopresentism, this requirement is violated, because on this view, there may be two events in the elsewhere region of here-now such that one is located at s and the other is located at s', where s PREC s' (see Fig. 9.5).

Savitt's requirement is formulated in terms of the notion of presentness; and his whole discussion is premised on the idea that relativistic counterparts to presentism must rehabilitate that notion within a relativistic setting. Once the revisionary strategy is adopted, this premise becomes somewhat contentious. However, Savitt's requirement is also presented as if it was a general principle not tied to any metaphysical theory. But then it should likewise be a plausible requirement for spatiopermanentists; and it is far from clear how this can be so, unless spatiopermanentists too are under the obligation to rehabilitate the notion of presentness. The only candidate notion of presentness, acceptable to the spatiopermanentist, would seem to be the notion of being located at the spacetime-point that is the referent of 'here-now'. But once presentness is understood along these lines, bow-tie spatiopresentism

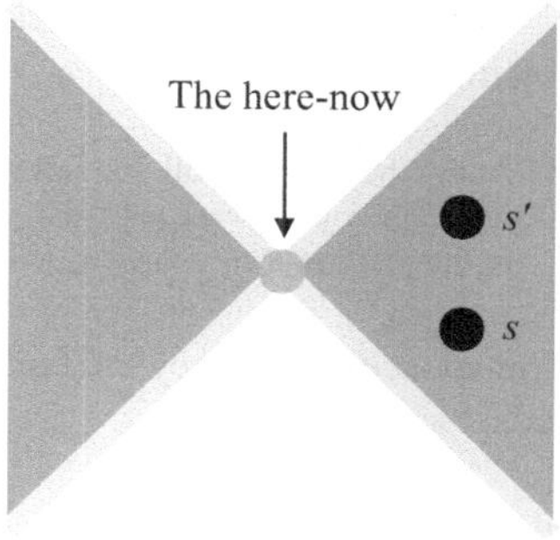

Fig. 9.5 Illustration of Savitt's objection

would seem to face no problem in its attempt to meet the requirement. Of course, one cannot then, at the same time, oblige proponents of bow-tie spatiopresentism to subscribe to the claim that, everywhere, whatever exists is present *in this sense*; for this would be tantamount to the flat-out assertion that spatiopresentism must be pointy.

Perhaps a better way with Savitt's objection would be to construe it as saying that spatiopresentists *qua* spatiopresentists should be committed to

(3) $\mathrm{E!}x \mathbin{\&} \mathrm{E!}y \mathbin{\&} \mathrm{S}x \mathbin{\&} \mathrm{S}y \rightarrow \neg(x \text{ PREC } y)$

It is clear that (3) is not theory-neutral, since spatiopermanentism and both versions of relativistic GBT reject it. Given its commitment to (1), pointy spatiopresentism obviously vindicates (3). The interesting question is whether spatiopresentists *qua* spatiopresentists are committed to (3). However, failing (PO), the *reductio* proof of (3), based on the two spatiopresentist principles ($\mathrm{P2_R}$) and ($\mathrm{P3_R}$) and mimicking the proofs of (1) and (2) as closely as possible, must at some stage appeal to the principle

(BO+) $\mathrm{E!}x \rightarrow \blacktriangleleft\mathrm{E!}x$[4]

which, given (AR1), (AR5) and (RR), is equivalent to '$\triangleleft\mathrm{E!}x \rightarrow \mathrm{E!}x$'.[5] However, (BO+) is much stronger than (BO), and so it is questionable why bow-tie spatiopresentists should be committed to it. We will comment on the implication of (BO+) further below, as (BO+) also makes its appearance in the context of another principle that has been used to discredit bow-tie views.

[4]Assume $\mathrm{E!}m \mathbin{\&} \mathrm{E!}n \mathbin{\&} \mathrm{S}m \mathbin{\&} \mathrm{S}n$. Assume for *reductio* m PREC n. By ($\mathrm{P2_R}$), ($\mathrm{P3_R}$), ($\mathrm{A34_R}$) and ($\mathrm{A23_R}$), this yields $@n(\mathrm{E!}n \mathbin{\&} \blacktriangle\neg\mathrm{E!}n \mathbin{\&} \blacktriangledown\neg\mathrm{E!}n \mathbin{\&} \neg\mathrm{E!}m)$. By ($\mathrm{T2_R}$), this yields $\bigcirc(\mathrm{E!}n \mathbin{\&} \blacktriangle\neg\mathrm{E!}n \mathbin{\&} \blacktriangledown\neg\mathrm{E!}n \mathbin{\&} \neg\mathrm{E!}m)$. Given our initial assumption, we can derive $\neg\triangle(\mathrm{E!}n \mathbin{\&} \blacktriangle\neg\mathrm{E!}n \mathbin{\&} \blacktriangledown\neg\mathrm{E!}n \mathbin{\&} \neg\mathrm{E!}m)$, $\neg\triangledown(\mathrm{E!}n \mathbin{\&} \blacktriangle\neg\mathrm{E!}n \mathbin{\&} \blacktriangledown\neg\mathrm{E!}n \mathbin{\&} \neg\mathrm{E!}m)$ and $\neg(\mathrm{E!}n \mathbin{\&} \blacktriangle\neg\mathrm{E!}n \mathbin{\&} \blacktriangledown\neg\mathrm{E!}n \mathbin{\&} \neg\mathrm{E!}m)$. This leaves as the only option $\triangleleft(\mathrm{E!}n \mathbin{\&} \blacktriangle\neg\mathrm{E!}n \mathbin{\&} \blacktriangledown\neg\mathrm{E!}n \mathbin{\&} \neg\mathrm{E!}m)$ which, it would seem, can only be rejected if (BO+) is assumed.

[5]Suppose $\varphi \rightarrow \blacktriangleleft\varphi$ is a theorem. By (AR1) and (RR), then, $\triangleleft\varphi \rightarrow \triangleleft\blacktriangleleft\varphi$ is a theorem. Using (AR5) and again (AR1) and (RR), $\triangleleft\blacktriangleleft\varphi \rightarrow \varphi$ is a theorem. So, $\triangleleft\varphi \rightarrow \varphi$ is a theorem. Conversely, suppose the latter is a theorem. Then by (AR1) and (RR), so is $\blacktriangleleft\triangleleft\varphi \rightarrow \blacktriangleleft\varphi$. By (AR5), then, $\varphi \rightarrow \blacktriangleleft\varphi$ is a theorem.

Thus, in his critique of presentism in the context of SR, Putnam invokes the following principle:

> If it is the case that all and only the things that stand in a certain relation *R* to me-now are real, and you-now are also real, then it is also the case that all and only the things that stand in the relation *R* to you-now are real. (Putnam 1967: 241)

Transposed to the key of our current discussion, the principle may faithfully be recast as.

(4) $\mathrm{E}!x \;\&\; @x\mathrm{E}!y \rightarrow \mathrm{E}!y$

Given spatiopermanentism, (4) trivially holds. The proof that (4) also holds on pointy spatiopresentism, and on pointy relativistic GBT, is straightforward:

Proof: Pointy spatiopresentism: Assume both $\mathrm{E}!x$ and $@x\mathrm{E}!y$. By (P2$_{\mathrm{R}}$), (P3$_{\mathrm{R}}$), (PO) and (A23$_{\mathrm{R}}$), the latter yields: $@x(\mathrm{E}!x \;\&\; \blacktriangle\neg\mathrm{E}!x \;\&\; \blacktriangledown\neg\mathrm{E}!x \;\&\; \blacktriangleleft\neg\mathrm{E}!x \;\&\; \mathrm{E}!y)$. By (T2$_{\mathrm{R}}$), $\bigcirc(\mathrm{E}!x \;\&\; \blacktriangle\neg\mathrm{E}!x \;\&\; \blacktriangledown\neg\mathrm{E}!x \;\&\; \blacktriangleleft\neg\mathrm{E}!x \;\&\; \mathrm{E}!y)$. Given $\mathrm{E}!x$, $\mathrm{E}!y$. Pointy relativistic GBT: Assume both $\mathrm{E}!x$ and $@x\mathrm{E}!y$. By (P2$_{\mathrm{R}}$), (PO) and (A23$_{\mathrm{R}}$), the latter yields: $@x(\mathrm{E}!x \;\&\; \blacktriangle\neg\mathrm{E}!x \;\&\; \blacktriangleleft\neg\mathrm{E}!x \;\&\; \mathrm{E}!y)$. By (T2$_{\mathrm{R}}$), $\bigcirc(\mathrm{E}!x \;\&\; \blacktriangle\neg\mathrm{E}!x \;\&\; \blacktriangleleft\neg\mathrm{E}!x \;\&\; \mathrm{E}!y)$. Given $\mathrm{E}!x$ and (P1$_{\mathrm{R}}$), $\mathrm{E}!y$. □

On bow-tie views, by contrast, (4) can only be derived in analogous ways on condition that (BO+) holds. Thus, assume bow-tie spatiopresentism and assume both $\mathrm{E}!x$ and $@x\mathrm{E}!y$. By (P2$_{\mathrm{R}}$), (P3$_{\mathrm{R}}$), (BO) and (A23$_{\mathrm{R}}$), the latter yields: $@x(\mathrm{E}!x \;\&\; \blacktriangle\neg\mathrm{E}!x \;\&\; \&\; \blacktriangledown\neg\mathrm{E}!x \;\&\; \blacktriangleleft\mathrm{E}!x \;\&\; \mathrm{E}!y)$. By (T2$_{\mathrm{R}}$), $\bigcirc(\mathrm{E}!x \;\&\; \blacktriangle\neg\mathrm{E}!x \;\&\; \blacktriangledown\neg\mathrm{E}!x \;\&\; \blacktriangleleft\mathrm{E}!x \;\&\; \mathrm{E}!y)$. Given $\mathrm{E}!x$, either $(\mathrm{E}!x \;\&\; \blacktriangle\neg\mathrm{E}!x \;\&\; \blacktriangledown\neg\mathrm{E}!x \;\&\; \blacktriangleleft\mathrm{E}!x \;\&\; \mathrm{E}!y)$ or $\triangleleft(\mathrm{E}!x \;\&\; \blacktriangle\neg\mathrm{E}!x \;\&\; \blacktriangledown\neg\mathrm{E}!x \;\&\; \blacktriangleleft\mathrm{E}!x \;\&\; \mathrm{E}!y)$. In the first case, $\mathrm{E}!y$. But in order to be able to derive $\mathrm{E}!y$ in the second case, we would need '$\triangleleft\mathrm{E}!y \rightarrow \mathrm{E}!y$', which is equivalent to (BO+).

Alternatively, assume bow-tie relativistic GBT and assume both $\mathrm{E}!x$ and $@x\mathrm{E}!y$. By (P2$_{\mathrm{R}}$), (BO) and (A23$_{\mathrm{R}}$), the latter yields: $@x(\mathrm{E}!x \;\&\; \blacktriangle\neg\mathrm{E}!x \;\&\; \blacktriangleleft\mathrm{E}!x \;\&\; \mathrm{E}!y)$. By (T2$_{\mathrm{R}}$), $\bigcirc(\mathrm{E}!x \;\&\; \blacktriangle\neg\mathrm{E}!x \;\&\; \blacktriangleleft\mathrm{E}!x \;\&\; \mathrm{E}!y)$. Given $\mathrm{E}!x$, either $(\mathrm{E}!x \;\&\; \blacktriangle\neg\mathrm{E}!x \;\&\; \blacktriangleleft\mathrm{E}!x \;\&\; \mathrm{E}!y)$ or $\triangle(\mathrm{E}!x \;\&\; \blacktriangle\neg\mathrm{E}!x \;\&\; \blacktriangleleft\mathrm{E}!x \;\&\; \mathrm{E}!y)$ or $\triangleleft(\mathrm{E}!x \;\&\; \blacktriangle\neg\mathrm{E}!x \;\&\; \blacktriangleleft\mathrm{E}!x \;\&\; \mathrm{E}!y)$. In the first two cases, given (P1$_{\mathrm{R}}$), $\mathrm{E}!y$. But in order to be able to derive $\mathrm{E}!y$ in the third case, we would need '$\triangleleft\mathrm{E}!y \rightarrow \mathrm{E}!y$', which is equivalent to (BO+).

Accordingly, like the (PO)-independent proof of (3) – the principle we extracted from Savitt (2000) – the (PO)-independent proof of (4) relies on

(BO+) $\mathrm{E}!x \rightarrow \blacktriangleleft\mathrm{E}!x$

But now, as long as we limit the range of viable views to views whose ontological commitments do not trace any more fine-grained divisions than those the fourfold causal structure of spacetime permits, (BO+) can easily be seen to collapse into (SPER). It is clear that (BO+) fails on the pointy views above – and similarly on the pointy view, not discussed here, according to which only the here-now and the causal future of here-now are populated: (BO+) together with any pointy view yields $@x\blacktriangleleft(\mathrm{E}!x \;\&\; \neg\mathrm{E}!x)$, which corresponds to the fact that the relation of causal precedence between spacetime-points is a total relation. Using ($A22_R$) it can indeed be shown that if we have @x◀(E!x & ¬E!x) as a theorem, then ¬◁φ is a theorem, and so, ($A5_R$) and ($A6_R$) boil down to the usual past- and future-linearity principles. The upshot is that, granted that we are in a genuinely relativistic setting, (BO+) fails on pointy views.

On bow-tie views, as Savitt notes, there are points s and s' in the elsewhere region of here-now such that s is in the causal future of s'. Accordingly, if at s, s did not exist anywhere in the causal past of s – as per ($P2_R$) it did not – then s cannot exist at s' – contrary to what (BO+) would demand. (BO+) is similarly at odds with the bow-tie view, not discussed here, according to which only the causal past is unpopulated: if at s', s' will not exist anywhere in the causal future of s', then s' cannot exist at s – contrary to what (BO+) demands. The only principled option left is spatiopermanentism.

Another way of arguing that, in a relativistic setting, (BO+) is in tension with spatiotemporaryism quite generally, is this. Consider the converse of (AR2), which is equivalent to ◀◀φ → ●φ. Given (BO+), and given that ◀ behaves normally, ◀E!x → ◀◀E!x holds, and hence, again by (BO+), so does E!x → ◀◀E!x. By the converse of (AR2), we infer E!x → ●E!x, which, by ($A16_R$) and the fact that ● is an S5 modality, yields ●E!x, and hence spatiopermanentism. In a relativistic setting, the converse of (AR2) looks very weak; it anyway holds in standard Minkowskian spacetime, and presumably in many 'well-behaved' spacetimes conforming to GR. Insofar as in combination with (BO+), the converse of (AR2) yields spatiopermanentism, spatiotemporaryists should refuse taking (BO+) on board.

Accordingly, pointy spatiopresentists can accept both (3) and (4), but must reject (BO+). Proponents of pointy relativistic GBT can accept (4), but must reject (3) and (BO+). Bow-tie spatiopresentists and proponents of bow-tie relativistic GBT must reject all of (3), (4) and (BO+). It is true that rejection of (4) implies a more thoroughgoing departure from the prerelativistic counterparts of either of the latter types of view. For, in a prerelativistic setting,

E!x & (At x, E!y) → E!y

which is the principle corresponding to (4), is easily proved to hold on any of the classical views.

Proof: (i) Permanentism: On permanentism, E!y anyway holds. (ii) GBT: Assume both E!x and At x, E!y. Given (P2) and (A23), the latter yields: At x, (E!x & H¬E!x & E!y). By (T2), from this we derive: Sometimes, (E!x & H¬E!x & E!y). Since E!x, either (E!x & H¬E!x & E!y) or P(E!x & H¬E!x & E!y), and hence either E!y or PE!y. Either way, given (P1), E!y follows. (iii) Presentism: Assume both E!x and At x, E!y. Given (P3), (P2) and (A23), the latter yields: At x, (E!x & H¬E!x & G¬E!x & E!y). By (T2), from this derive: Sometimes, (E!x & H¬E!x & G¬E!x & E!y). Since E!x, E!y follows. □

Similarly, classical presentism vindicates

$$\mathrm{E}!x \,\&\, \mathrm{E}!y \,\&\, \mathrm{T}x \,\&\, \mathrm{T}y \rightarrow \neg(x \prec y)$$

which corresponds to (3). The only candidate prerelativistic principle corresponding to (BO+) would be the present-tensed tautology 'E!x → E!x', if only because, in the prerelativistic setting, all non-causally separated points, or events, are contemporaneous. To the extent that it is uncontroversial that in a relativistic setting, all spatiotemporaryist views must reject (BO+) and there is no objective sense in which non-causally separated points, or events, are contemporaneous, it should not be surprising that some of the counterparts of other prerelativistic principles must, in such a setting, likewise be given up – especially if, as argued here, vindication of those counterpart principles would, within the context of certain theoretical choices, require acceptance of (BO+).

Accordingly, it is far from clear why bow-tie spatiotemporaryism's rejection of (4) should be held against it, solely on the grounds that its prerelativistic precursors adopted its prerelativistic counterpart. Again, the appeal to everyday intuition has little to recommend itself, if only because the idea that there is a region of points non-causally separated from here-now some of which stand in relations of causal precedence to one another is itself not very intuitive by everyday standards.

We conclude that the bow-tie versions of spatiotemporaryism – not only are superior to their pointy rivals, in that unlike the latter, they can account for the persistence of non-causally separated objects of which we have perceptual knowledge – but also prove quite resilient in the face of prominent, principle-driven attempts to discredit them, as no convincing case has been made for thinking that spatiotemporaryists must show allegiance to the principles invoked. That bow-tie versions of spatiotemporaryism are the most natural candidates for serving as the relativistic counterparts to classical presentism and classical GBT, respectively, will independently be confirmed in the next and last section of this chapter.

9.5 Recapture of the Classical Views

We have hitherto been concerned with the articulation of relativistic theories of spacetime that promise to be natural counterparts of prerelativistic theories of time. However, so far, we have left the relation between relativistic theories and their prerelativistic counterparts at a somewhat intuitive level. As advertised towards the end of Chap. 8, to complete implementation of the revisionary strategy, we must find a principle, or conjunction of principles, q, couched in relativistic terms, such that for any prerelativistic theory T, its relativistic counterpart T_R implies T on condition that q holds, where q, though consistent with T_R, is inconsistent with rejection of an absolute and total temporal order. We may then conceive of T as being decomposable into two independent parts, q and T_R, such that even if q fails, T_R may nonetheless survive relativity.

The distinction between pointy and bow-tie versions of spatiotemporaryism might suggest that one such prerelativistic theory *T* can have more than one relativistic counterpart, so that the decomposition might not be unique, which is unproblematical as long as, given *q*, these relativistic counterparts collapse into one another. However, as we shall see, such a collapse is not to be had and bow-tie views prove the by far more natural candidates for being relativistic counterparts of classical versions of temporaryism.

The search for such a *q* must be guided by the following thought. Prerelativistic physics presumes that there are spatially separated events that are absolutely simultaneous with one another, and so in effect, that there are distinct spacetime-points which share the same absolute time coordinate. The principle *q* should accordingly allow us to conceive of absolute times as corresponding to equivalence classes of spacetime-points such that anywhere any spacetime-point non-causally separated from, or identical with, some member of such a class *eo ipso* belongs to that same class. This *inter alia* requires that, given *q*, the relation SIM should prove to be an equivalence relation:

$$x \text{ SIM } y \equiv_{df} Sx \,\&\, Sy \,\&\, \neg(x \text{ PREC } y) \,\&\, \neg(y \text{ PREC } x)$$

Similarly, *q* should allow us to conceive of relations of precedence among absolute times as corresponding to suitable relations among such equivalence classes, and of relations of absolute temporal location in terms of suitable relations of being somewhere located at some member of such a class.

These are not the only constraints on *q* that should guide us, however. In prerelativistic theories of time, with their absolute conception of tense, what is the case at most varies across time, but not across space. Accordingly, a further thought must be that, given *q*, whatever is the case at any one member of such an equivalence class, likewise holds at any other member of that class.

In this spirit, we now suggest that the sought-after *q* should be conceived of as the conjunction of the following two bridge principles:

(PR1) $x \text{ PREC } y \,\&\, x \text{ SIM } z \rightarrow z \text{ PREC } y$
(PR2) $@x\varphi \,\&\, x \text{ SIM } y \rightarrow @y\varphi$

In Appendix 3, we shall provide all the nitty-gritty details of how the combination of these bridge principles successfully serves its purpose. Here, we merely sketch some central features. Thus, note that (PR1) straightforwardly implies that SIM is transitive. Consequently, the bow-tie region of each spacetime-point, comprising that point and its elsewhere region, shrinks, as it were, to a single hypersurface. Given that we anyway have both '$Sx \rightarrow x \text{ SIM } x$' and '$x \text{ SIM } y \rightarrow y \text{ SIM } x$', (PR1) accordingly ensures that SIM is an equivalence relation, which is just as needed. It should also be clear that in a relativistic setting, (PR1) fails, precisely because the bow-tie region then has depth.

(PR2) expresses that there is no factual variation across members of the same equivalence class of spacetime-points, which is also just as needed. (PR2) implies both of the following:

(TR1) $\lhd\varphi \to \varphi$
(TR2) $\varphi \,\&\, \lhd\top \to \lhd\varphi$ (with $\top$ an arbitrary tautology)

Proof of (TR1): Let x be a variable that is not free in φ. From (AR6), i.e.

$$@x\blacktriangleleft\varphi \leftrightarrow Sx \,\&\, \bullet\forall y(y \text{ SEP } x \to @y\varphi)$$

we get $@x\lhd\varphi \to \bigcirc\exists y(y \text{ SEP } x \,\&\, @y\varphi)$. Given (PR2) we then get $@x\lhd\varphi \to @x\varphi$, and then $Sx \to @x(\lhd\varphi \to \varphi)$, and then $\forall x(Sx \to @x(\lhd\varphi \to \varphi))$. Using ($A22_R$), i.e.

$$\forall x(Sx \to @x\varphi) \to \varphi \quad \text{(with } x \text{ not free in } \varphi)$$

one can then derive $\lhd\varphi \to \varphi$. □

Proof of (TR2): From (TR1), we get $\varphi \to \blacktriangleleft\varphi$, and hence $\varphi \to \blacktriangleleft(\top \to \varphi)$. (TR2) follows, since we have in general $\blacktriangleleft(\psi \to \xi) \to (\lhd\psi \to \lhd\xi)$. □

It should be clear that, at least from a spatiotemporaryist perspective, neither (TR1) nor (TR2) holds in a relativistic setting. By contrast, in a prerelativistic setting – where the bow-tie region of each spacetime-point shrinks to a single hypersurface, and where there is assumed to be no factual variation across points on such a hypersurface – both (TR1) and (TR2) should hold.

In order to derive prerelativistic views from their relativistic counterparts *modulo* our bridge principles (PR1) and (PR2), we first systematically enrich the relativistic spacetime-point-sensitive language by a dyadic predicate ε for class membership, as well as a denumerably infinite stock of variables distinct from the variables of the original language. Since we are only interested in classes of spacetime-points, ε is understood to be restricted to membership in such classes. We adopt the following definition of the predicate C for classes:

$$Cm \equiv_{df} \bigcirc\exists x(x \varepsilon m) \quad \text{(with } x \text{ the first new variable distinct from } m)$$

Of all the axioms governing ε which we mention in Appendix 3, we here merely highlight the following comprehension axiom:

(AX6) $Sx \to @x\exists y\bullet\forall z(z \,\varepsilon\, y \leftrightarrow z \text{ SIM } x \,\&\, @xE!z))$

We next define:

– $SCm \equiv_{df} Cm \,\&\, \bullet\forall x\bullet\forall y(x \,\varepsilon\, m \,\&\, y \,\varepsilon\, m \to x \text{ SIM } y)$
(with x and y respectively the first and second new variable distinct from m)

– $m \text{ SIMEM } n \equiv_{df} \bigcirc\exists x(m \text{ SIM } x \,\&\, x \,\varepsilon\, n)$
(with x the first new variable distinct from both m and n)

Here 'SC' can be read as being short for 'is a simultaneity class' and 'SIMEM' as being short for 'is simultaneous with a member of'.

Armed with these definitions, and on assumption of (PR1) and (PR2), we can now define notions that respectively correspond to the prerelativistic notions of a time-instant, temporal precedence among such time-instants, location at a time-instant, and holding at a time-instant:

- $\mathrm{T}^*m \equiv_{\mathrm{df}} \mathrm{SC}m \;\&\; \bullet\forall x(x \text{ SIMEM } m \rightarrow x \,\varepsilon\, m)$
- $m \prec^* n \equiv_{\mathrm{df}} \mathrm{T}^*m \;\&\; \mathrm{T}^*n \;\&\; \bullet\forall x \bullet\forall y(x \,\varepsilon\, m \;\&\; y \,\varepsilon\, n \rightarrow x \text{ PREC } y)$
- $m\, \mathrm{L}^*\, n \equiv_{\mathrm{df}} \mathrm{T}^*n \;\&\; \circ\exists x(x \,\varepsilon\, n \;\&\; m \text{ LOC } x)$
- $\mathrm{At}^*\, m, \varphi \equiv_{\mathrm{df}} \mathrm{T}^*m \;\&\; \bullet\forall x(x \,\varepsilon\, m \rightarrow @x\varphi)$

In the presence of (PR1) and (PR2), T*, ≺*, L* and At* can indeed serve as relativistic translations of the prerelativistic T, ≺, L and At. A detailed translation manual taking the prerelativistic language into the relativistic language can be found in Appendix 3. However, we note already here that ▲ and ▼ serve as natural translations of H and G, respectively.

We can now establish the following theorems:

(A30#) $\neg(x \prec^* x)$
(A31#) $(x \prec^* y \;\&\; y \prec^* z) \rightarrow x \prec^* z$
(A32#) $(\mathrm{T}^*x \;\&\; \mathrm{T}^*y) \rightarrow (x \prec^* y \vee x = y \vee y \prec^* x)$

(A30#) already follows from the aforementioned definitions, while (A31#) can be established using ($A31_R$), i.e. '$(x \text{ PREC } y \;\&\; y \text{ PREC } z) \rightarrow x \text{ PREC } z$'. The totality principle (A32#), by contrast, depends on (PR1). (A30#) to (A32#) mimic the classical axioms (A30) to (A32) which latter encode central features of the absolute and total temporal order that is characteristic of prerelativistic physics.

The systematic recapture of the other prerelativistic axioms, including those for temporal location, can be found in Appendix 3. There, we will give a translation of the prerelativistic language into the enriched relativistic language and show in detail that in the presence of our bridge principles, for any theorem of the classical neutral system described in Appendix 1, its translation is a theorem of the relativistic neutral system described in Appendix 2, and for any theorem of classical GBT/presentism/permanentism, its translation is a theorem of relativistic GBT/presentism/permanentism.

Before closing, let us here briefly review how the more specific principles introduced in the previous section fare, once (PR1) and (PR2) are added to the stock. We begin by considering the characteristic principles for pointy and for bow-tie views, i.e.

(PO) $\mathrm{S}x \rightarrow @x\blacktriangleleft\neg \mathrm{E!}x$
(BO) $\mathrm{S}x \rightarrow @x\blacktriangleleft \mathrm{E!}x$

(BO) already follows from (TR1) and ($A23_R$), i.e. '$\mathrm{S}x \rightarrow @x\mathrm{E!}x$'. If (PO) is accepted, then given (BO) and (AR6), one can derive:

(2) $\mathrm{S}x \rightarrow \bullet\neg\exists y(x \text{ SEP } y)$

which latter is equivalent to:

– $Sx \rightarrow \bullet\forall y(x \text{ SIM } y \rightarrow x = y)$

This shows very clearly that pointy views are implausible candidates for being the relativistic counterparts of classical temporaryist views: there is no suggestion in classical presentism that space shrinks to a single point, and no suggestion in classical GBT either that the edge of becoming is not really an edge but a point. This leaves bow-tie spatiopresentism and bow-tie relativistic GBT as the only serious contenders. As argued in the previous section, in a relativistic setting, both these views must reject.

(BO+) $E!x \rightarrow \blacktriangleleft E!x$

By contrast, we argued that in such a relativistic setting, the converse of (AR2), i.e.

(C-AR2) $\bigcirc\varphi \rightarrow \triangleleft\triangleleft\varphi$

is eminently plausible. This assessment is reversed in a prerelativistic setting. Thus, once (PR2) is added to the mix, (TR1) becomes derivable, and (TR1) straightforwardly implies (BO+). However, given that same theorem (TR1), (C-AR2) straightforwardly implies $\bigcirc\varphi \rightarrow \varphi$, and so in particular, $\vartriangle\varphi \rightarrow \varphi$ and $\triangledown\varphi \rightarrow \varphi$. None of these latter principles is acceptable to spatiotemporaryists, irrespective of whether (PR1) and (PR2) are assumed to hold.

Rejecting (C-AR2) while accepting (BO+) thus becomes mandatory for spatiotemporaryists, whether they defend pointy or bow-tie views, once we change from a relativistic to a prerelativistic setting, in which latter setting both our bridge principles hold. Note that this reversal in no way undermines the contention that, in the presence of (PR1) and (PR2), for any theorem of prerelativistic views, its relativistic translation is a theorem of their respective relativistic counterparts. For, neither (C-AR2) nor the negation of (BO+) is a theorem of the relevant relativistic views taken on their own, i.e. without making further assumptions about whether or not we find ourselves in a relativistic setting in which the elsewhere region has depth.

We similarly noted in the previous section that in a relativistic setting, bow-tie spatiopresentists and proponents of bow-tie relativistic GBT alike must reject both of the following:

(3) $E!x \,\&\, E!y \,\&\, Sx \,\&\, Sy \rightarrow \neg(x \text{ PREC } y)$
(4) $E!x \,\&\, @xE!y \rightarrow E!y$

In a prerelativistic setting, however, in which both (PR1) and (PR2) hold, bow-tie spatiopresentists will accept both (3) and (4), alongside their pointy friends, while proponents of bow-tie relativistic GBT, just like *their* pointy friends, will continue rejecting (3), while now accepting (4). Again, these findings in no way challenge our claim that, in the presence of (PR1) and (PR2), for any theorem of prerelativistic views, its relativistic translation is a theorem of their respective relativistic counterparts.

This concludes our discussion of the revisionary strategy and the metaphysical views its implementation allows us to articulate (however, see Appendix 2 and Appendix 3 for a more thoroughgoing formal treatment). In the previous chapter we conceded to the friends of the conservative strategy that it is an open empirical question whether physics may ultimately reinstate assumption of an absolute and total temporal order. So, in the end there may prove to be no need to contemplate the different spatiotemporaryist views expounded here. However, it was part of the rationale of the revisionary strategy that we should not make our metaphysical options hostage to such empirical fortune and prepare for the case in which the relativistic conception of spacetime persists. Although we do not pretend to have given knock-down arguments against pointy versions of spatiotemporaryism, and have not even begun to negotiate between the two bow-tie versions introduced above, the mere provision of a theoretical and logical framework in which to articulate and discuss such issues seems to us to be a major advance in the metaphysical study of (space)time.

Concluding Remarks

Our main aim was to give a limited defence of GBT by showing that there is a coherent, logically perspicuous and ideologically lean formulation of it that is suited to successfully rebuff extant philosophical arguments against its cogency.

In particular, we argued that properly understood, GBT allows for (1) unrestricted quantification and an ontology of 3D as well as 4D objects, including their sets and mereological fusions, (2) easy and continually available knowledge of where the block's edge is located, (3) a coherent conception of purely past things as belonging to familiar kinds, and as having lost some of their tensed properties in virtue of new additions to the block – which should suffice to show that the view is not some ill-conceived hybrid between presentism about the future and eternalism about the past, (4) the bivalence of future contingents and a grounding requirement appropriate for tensed truths, (5) a strong indeterminist conception of the open future and the fixed past unavailable to its competitors, and (6) its systematic recapture in relativistically acceptable terms, preserving much of the spirit of the original.

By using limited conceptual resources, we have also shown (7) that GBT has no need for the properties of being present, past or future, or a duplication of tenses or time-dimensions, or any other exotic and unheard-of conceptual machinery – save perhaps in the context of relativistic physics where a spacetime-sensitive language and logic would anyway seem to be needed. Along the way, we have also made clear (8) that GBT can easily accommodate the standard semantics for token-reflexive expressions as well as Evans' observations concerning the absolute truth of token-utterances involving such expressions, thereby averting the threat of radical relativism.

Unlike Tooley's version of the view, our version (9) successfully averts hostile takeover by static theories of time, in that certain of its key claims openly contradict those made by permanentists.

We have furthermore shown (10) that being a version of temporaryism, GBT can fruitfully be discussed in the terms that Timothy Williamson has recently proposed as being best suited for conducting ontological debates about temporal existence.

F. Correia, S. Rosenkranz, *Nothing To Come*, Synthese Library 395,
https://doi.org/10.1007/978-3-319-78704-6

Ten points that we think should suffice for putting GBT back on the agenda. These results can be appreciated even by those who harbour no sympathies for it. If intent on showing that GBT is inferior to their own preferred view, they now need to devise better arguments for rejecting it. Appeals to the epistemic objection, complaints about the weird status of past objects, doubts about the conceptual coherence of devices such as those used by Tooley and Button, and the charge that GBT is eternalism is disguise, will simply no longer do.

At the same time, however, proponents of presentism and dynamic permanentism can benefit from the logico-conceptual framework here provided. For, we have likewise shown that it affords formulations of those rival views that no longer rely on presentness, concreteness, ages, temporal distributional properties, or other props. Presentists, in particular, can readily avail themselves of the outcome of our discussion of truth-grounding and put it to use in defending their own view, without the need to invoke Lucretian properties or facts that invite the charge of cheating.

At last, the relativistic counterparts we introduced in Chap. 9, alongside a spatiotemporal logic and semantics suited for their discussion, open up interesting new avenues towards a reassessment of the relation between physics and metaphysics. As such, they should also be of concern to those scientifically minded philosophers who have for long dismissed dynamic views for their apparent inability to come to terms with relativity. There may of course be other reasons, furnished by physical theory, for rejecting such views; but in any case, it will no longer do to reject them simply by appeal to the structure of relativistic spacetime and its lack of provision of an absolute and total temporal order.

Appendix 1
Semantic Characterisation of the Classical Systems

The formal systems for classical GBT, presentism and permanentism introduced in previous chapters share core postulates, which define what we shall here call *the neutral system*, and differ by the addition of characteristic axioms – (P1) and (P2) for GBT, (P2) and (P3) for presentism, and (PER′) for permanentism. In this appendix, we review these four systems, and semantically characterise each of them in a homogeneous framework.

1. The languages

Each system can be formulated in a language whose vocabulary is specified as follows:

- a countable stock of predicates, including the 2-place identity predicate =, the 1-place predicate T for times, the 2-place predicate $\prec$ for temporal precedence and the 2-place predicate L for location
- a countably infinite set of variables, for which we shall use x, y, etc.
- a countable (possibly empty) set of constants
- the truth-functional connectives $\neg$ and & and the quantifier $\forall$
- the Priorean operators H and G
- the temporal prenective At
- the brackets (and)

The *terms* of such a language, for which we shall use m, n, etc., are its variables and constants, and its *formulas* are recursively defined as follows:

- If Φ is an n-place predicate and $m_1, \ldots, m_n$ are terms, then $\Phi m_1 \ldots m_n$ is a formula.
- If φ is a formula, then so are $\neg\varphi$, Hφ and Gφ.
- If φ and ψ are formulas, then so is $(\varphi \,\&\, \psi)$.
- If φ is a formula and x a variable, then $\forall x\varphi$ is a formula.
- If φ is a formula and m a term, then At m, φ is a formula.

F. Correia, S. Rosenkranz, *Nothing To Come*, Synthese Library 395,
https://doi.org/10.1007/978-3-319-78704-6

Given any language defined in the specified way, standard conventions and notions are taken for granted, and the following definitions are adopted:

- $P\varphi \equiv_{df} \neg H \neg \varphi$
- $F\varphi \equiv_{df} \neg G \neg \varphi$
- Always, $\varphi \equiv_{df} H\varphi \mathbin{\&} \varphi \mathbin{\&} G\varphi$
- Sometimes, $\varphi \equiv_{df} P\varphi \vee \varphi \vee F\varphi$
- $E!m \equiv_{df} \exists x(x = m)$
  (where x is the first standard variable distinct from m given an established numbering of these)

The previous specification characterises, not one language, but a family of languages with different sets of constants or predicates distinct from =, T, < and L (or both). Up to section 6, we will suppose as given a fixed language £ of the sort just defined.

2. The neutral system

The neutral system is defined by the following axioms and rules (see Chaps. 1 to 3).

Classical propositional axioms and rule:

The rule Modus Ponens, plus any suitable set of axioms for classical propositional logic, for instance the familiar set proposed by Jan Łukasiewicz.

Axioms and rule for quantification and identity:

(A17) $\forall x(\varphi \to \psi) \to (\forall x\varphi \to \forall x\psi)$
(A18) $\varphi \to \forall x\varphi$
(A19) $\forall x\varphi \mathbin{\&} E!m \to \varphi[m/x]$
(A15) $\forall x E!x$
(A16) Sometimes, $E!m$
(R3) $\varphi \ / \ \forall x\varphi$

(A20) $x = x$
(A21) $x = y \to (\varphi \to \varphi[y//x])$

In (A18), x is not free in φ. In (A19), $\varphi[m/x]$ is the result of freely replacing each free occurrence of x in φ by m. In (A21), $\varphi[y//x]$ is the result of freely replacing zero or more free occurrences of x in φ by y. Axioms for T:

(A9) $Tx \to$ Always, Tx
(A23) $Tx \to$ At x, $E!x$

Axioms for $\prec$:

(A28) $x \prec y \rightarrow (\mathrm{T}x \;\&\; \mathrm{T}y)$
(A29) $x \prec y \rightarrow$ Always, $(x \prec y)$
(A30) $\neg(x \prec x)$
(A31) $(x \prec y \;\&\; y \prec z) \rightarrow x \prec z$
(A32) $(\mathrm{T}x \;\&\; \mathrm{T}y) \rightarrow (x \prec y \vee x = y \vee y \prec x)$

Axioms for L:

(A25) $x \mathrm{L} y \rightarrow \mathrm{T}y$
(A26) $\mathrm{T}x \rightarrow (x \mathrm{L} y \leftrightarrow x = y)$
(A27) $x \mathrm{L} y \rightarrow$ Always, $x \mathrm{L} y$

Axioms and rules for the Priorean operators:

(A1) $\varphi \rightarrow \mathrm{HF}\varphi$
(A2) $\varphi \rightarrow \mathrm{GP}\varphi$
(A3) $\mathrm{H}(\varphi \rightarrow \psi) \rightarrow (\mathrm{H}\varphi \rightarrow \mathrm{H}\psi)$
(A4) $\mathrm{G}(\varphi \rightarrow \psi) \rightarrow (\mathrm{G}\varphi \rightarrow \mathrm{G}\psi)$
(A5) $\mathrm{FP}\varphi \rightarrow (\mathrm{P}\varphi \vee \varphi \vee \mathrm{F}\varphi)$
(A6) $\mathrm{PF}\varphi \rightarrow (\mathrm{P}\varphi \vee \varphi \vee \mathrm{F}\varphi)$
(A7) $\mathrm{PP}\varphi \rightarrow \mathrm{P}\varphi$
(A14) $\mathrm{P}\top \vee \mathrm{F}\top$, for $\top$ any chosen tautology
(R1) $\varphi \;/\; \mathrm{H}\varphi$
(R2) $\varphi \;/\; \mathrm{G}\varphi$

Axioms for 'At':

(A8) (At x, φ) $\rightarrow \mathrm{T}x$
(A10) At x, $(\varphi \rightarrow \psi) \rightarrow$ (At x, $\varphi \rightarrow$ At x, ψ)
(A11) At x, $\neg\varphi \leftrightarrow (\mathrm{T}x \;\&\; \neg$At x, $\varphi)$
(A12) (Always, φ) $\rightarrow (\mathrm{T}x \rightarrow$ At x, φ)
(A13) (At x, φ) $\rightarrow$ Always, At x, φ
(A22) $\forall x(\mathrm{T}x \rightarrow$ At x, $\varphi) \rightarrow \varphi$
(A33) At x, $\mathrm{H}\varphi \leftrightarrow \mathrm{T}x$ & Always, $\forall y(y \prec x \rightarrow$ At y, φ)
(A34) At x, $\mathrm{G}\varphi \leftrightarrow \mathrm{T}x$ & Always, $\forall y(x \prec y \rightarrow$ At y, φ)
(A24) At x, $\forall y\varphi \leftrightarrow$ Always, $\forall y$ At x, (E!$y \rightarrow \varphi$)

In (A22), x is not free in φ. In (A33) and (A34), y is not free in φ. In (A33), (A34) and (A24), x and y must be distinct variables.

(Note that the set of axioms for 'At' could be somewhat simplified. For instance, given (A8), (A11) could be replaced by the pair of axioms 'At x, $\neg\varphi \rightarrow \neg$At x, φ' and '($\mathrm{T}x$ & $\neg$At x, φ) $\rightarrow$ At x, $\neg\varphi$', and the left-to-right direction of (A24) could be omitted since it can be derived using (A8), (A10), (A12), (A13) and postulates for quantification and the Priorean operators.)

3. The systems for GBT, presentism and permanentism

Each system is an extension of the neutral system obtained by adding one or two axioms (see Chap. 4 for GBT, and Chap. 5 for presentism and permanentism).

Characteristic axioms of GBT:

(P1) $\text{E!}x \rightarrow \text{GE!}x$
(P2) $\text{T}x \rightarrow \text{At } x, \text{H}\neg\text{E!}x$

Characteristic axioms of presentism:

(P2) $\text{T}x \rightarrow \text{At } x, \text{H}\neg\text{E!}x$
(P3) $\text{T}x \rightarrow \text{At } x, \text{G}\neg\text{E!}x$

Characteristic axiom of permanentism:

(PER′) $\text{E!}x$

4. Semantics

We define a *neutral model* for language £ as a tuple ⟨ Ti, Bef, D, Loc, I ⟩, where:

- Ti (times) is a set with at least two elements
- Bef (before) is a binary relation on Ti which is irreflexive, transitive and total
- D (domain) is a function taking each time u in Ti into a set of objects D(u), such that for all u in Ti, $u \in \text{D}(u)$
- Loc (localisation) is a function taking each time u in Ti into a set of objects Loc(u), such that for all $u, v \in \text{Ti}$, $u \in \text{Loc}(v)$ iff $u = v$
- I (interpretation) is a function which takes each constant into an element of $\cup_{u \in \text{Ti}} \text{D}(u)$, and each k-place predicate distinct from =, T, ≺ and L and each time u into a set of k-tuples of objects taken from $\cup_{u \in \text{Ti}} \text{D}(u)$

The models for GBT are the neutral models that satisfy the following conditions:

(p1) For all u, v with u Bef v, $\text{D}(u) \subseteq \text{D}(v)$
(As time goes by, the ontology never shrinks)
(p2) For all u, v with v Bef u, $u \notin \text{D}(v)$
(At any time, no later times exist)

The models for presentism are those that satisfy the following conditions:

(p2) For all u, v with v Bef u, $u \notin \text{D}(v)$
(At any time, no later times exist)
(p3) For all u, v with u Bef v, $u \notin \text{D}(v)$
(At any time, no previous times exist)

And the models for permanentism are those that satisfy the following condition:

(per′) For all $o \in \cup_{u \in \mathrm{Ti}} \mathrm{D}(u)$ and $u \in \mathrm{Ti}$, $o \in \mathrm{D}(u)$
(The ontology never varies across time)

The truth-conditions for the formulas in a neutral model $\langle$ Ti, Bef, D, Loc, I $\rangle$ at a time u relative to a variable-assignment r are as follows ($\mathrm{I}_r(m)$ is $\mathrm{I}(m)$ if m is a constant, and $r(m)$ if m is a variable):

[L1] For Φ a k-place predicate distinct from $=$, T, $\prec$ and L:
$u \vDash_r \Phi m_1 \ldots m_k$ iff $\langle \mathrm{I}_r(m_1), \ldots, \mathrm{I}_r(m_k) \rangle \in \mathrm{I}(\Phi, u)$
[L2] $u \vDash_r m = m^*$ iff $\mathrm{I}_r(m)$ is $\mathrm{I}_r(m^*)$
[L3] $u \vDash_r \mathrm{T}m$ iff $\mathrm{I}_r(m) \in \mathrm{Ti}$
[L4] $u \vDash_r m \prec m^*$ iff $\mathrm{I}_r(m)$ Bef $\mathrm{I}_r(m^*)$
[L5] $u \vDash_r m \mathrm{\,L\,} m^*$ iff $\mathrm{I}_r(m) \in \mathrm{Loc}(\mathrm{I}_r(m^*))$
[L6] $u \vDash_r \neg\varphi$ iff $u \nvDash_r \varphi$
[L7] $u \vDash_r \varphi \,\&\, \psi$ iff both $u \vDash_r \varphi$ and $u \vDash_r \psi$
[L8] $u \vDash_r \forall x\varphi$ iff $u \vDash_s \varphi$ for all assignments s differing from r at most on x and such that $s(x) \in \mathrm{D}(u)$
[L9] $u \vDash_r \mathrm{H}\varphi$ iff $v \vDash_r \varphi$ for all v such that v Bef u
[L10] $u \vDash_r \mathrm{G}\varphi$ iff $v \vDash_r \varphi$ for all v such that u Bef v
[L11] $u \vDash_r \mathrm{At}\ m, \varphi$ iff $\mathrm{I}_r(m) \vDash_r \varphi$

Note that we have, in particular,

- $u \vDash_r \mathrm{E!}m, \varphi$ iff $\mathrm{I}_r(m) \in \mathrm{D}(u)$

A formula is said to *hold* in a model iff it is true at all times of the model relative to all variable-assignments.

5. Soundness of the systems

Four systems have been defined, the neutral system and the systems for GBT, presentism and permanentism. Each system is sound with respect to its associated semantics:

Theorem 1 (Soundness of the four systems). Every theorem of the neutral / GBT / presentist / permanentist system holds in all neutral / GBT / presentist / permanentist models.

Proof: (i) It is routine to verify that the four rules, which are common to the four systems, namely Modus Ponens, (R1), (R2) and (R3), preserve the property of holding in a neutral model. (ii) It is also routine to verify that the axioms for classical propositional logic, quantification and identity and the Priorean operators hold in every neutral model (and hence in every model for GBT, presentism or permanentism). The remaining neutral axioms involve vocabulary which is foreign to standard quantified temporal logic. Some can swiftly be seen to hold in every neutral model, for some others this requires a bit of work. (iii) Clearly, (P1) holds in every neutral model satisfying (p1), (P2) in every neutral model satisfying (p2), (P3) in every neutral model satisfying (p3), and (PER′) in every neutral model satisfying (per′). The details are left to the reader. □

6. Completeness of the neutral system

We consider a language £+, which is our original language £ plus a countably infinite set C+ of new constants, and we assume as given a numbering of the constants in C+. The systems were defined in, and the semantics was defined for, language £, but of course the definitions extend to £+.

Where S is a system extending the neutral system, we say that a set Δ of formulas of £ or £+ is S-*consistent* iff there are no formulas $\varphi_1, \ldots, \varphi_n \in \Delta$ of the language such that $\neg(\varphi_1 \,\&\, \ldots \,\&\, \varphi_n)$ is a theorem of S.

Say that a set Δ of £+-formulas is *nice* relative to a system S extending the neutral system iff:

- Δ is £+-maximal, i.e. for all £+-formulas φ, either $\varphi \in \Delta$ or $\neg\varphi \in \Delta$
- Δ is S-consistent
- Δ is ∀-saturated, i.e. the £+-formula ∀*x*φ belongs to Δ provided that E!*m* → φ[*m*/*x*] does for all £+-constants *m*
- Δ is H-∀-saturated, i.e. the £+-formula H∀*x*φ belongs to Δ provided that H(E!*m* → φ[*m*/*x*]) does for all £+-constants *m*
- Δ is G-∀-saturated, i.e. the £+-formula G∀*x*φ belongs to Δ provided that G(E!*m* → φ[*m*/*x*]) does for all £+-constants *m*
- Δ is At-∀-saturated, i.e. the £+-formula At *n*, ∀*x*φ belongs to Δ provided that At *n*, (E!*m* → φ[*m*/*x*]) does for all £+-constants *m*

Note that if Δ is nice, then:

- Δ is Always-∀-saturated, i.e. the £+-formula Always, ∀*x*φ belongs to Δ provided that Always, (E!*m* → φ[*m*/*x*]) does for all £+-constants *m*

and also:

- Δ is ∃-saturated, i.e. if the £+-formula ∃*x*φ belongs to Δ, then so does the formula E!*m* & φ[*m*/*x*] for some £+-constant *m*
- Δ is P-∃-saturated, i.e. if the £+-formula P∃*x*φ belongs to Δ, then so does the formula P(E!*m* & φ[*m*/*x*]) for some £+-constant *m*
- Δ is F-∃-saturated, i.e. if the £+-formula F∃*x*φ belongs to Δ, then so does the formula F(E!*m* & φ[*m*/*x*]) for some £+-constant *m*
- Δ is At-∃-saturated, i.e. if the £+-formula At *n*, ∃*x*φ belongs to Δ, then so does the formula At *n*, (E!*m* & φ[*m*/*x*]) for some £+-constant *m*
- Δ is Sometimes-∃-saturated, i.e. if the £+-formula Sometimes, ∃*x*φ belongs to Δ, then so does the formula Sometimes, (E!*m* & φ[*m*/x]) for some £+-constant *m*

Let S be a system that extends the neutral system, and let Δ be a set of £+-formulas that is nice relative to S. Being £+-maximal and S-consistent, Δ contains every theorem of S, and Δ is closed under detachment, i.e. if both φ and $\varphi \rightarrow \psi$ are in Δ, then so is ψ. We have in addition:

- $\neg\varphi \in \Delta$ iff $\varphi \notin \Delta$
- $\varphi \,\&\, \psi \in \Delta$ iff both $\varphi \in \Delta$ and $\psi \in \Delta$

The proof of Lemma 1 below will make use of the following four facts:

(a) If $\psi \rightarrow (E!m \rightarrow \varphi[m/x])$ is a theorem of S, where m is a constant that appears neither in φ nor in ψ, then $\psi \rightarrow \forall x\varphi$ is also a theorem of S
(b) If $\psi \rightarrow H(E!m \rightarrow \varphi[m/x])$ is a theorem of S, where m is a constant that appears neither in φ nor in ψ, then $\psi \rightarrow H\forall x\varphi$ is also a theorem of S
(c) If $\psi \rightarrow G(E!m \rightarrow \varphi[m/x])$ is a theorem of S, where m is a constant that appears neither in φ nor in ψ, then $\psi \rightarrow G\forall x\varphi$ is also a theorem of S
(d) If $\psi \rightarrow \text{At } n, (E!m \rightarrow \varphi[m/x])$ is a theorem of S, where m is a constant distinct from n that appears neither in φ nor in ψ, then $\psi \rightarrow \text{At } n, \forall x\varphi$ is also a theorem of S

Proof: For (a), suppose $\psi \rightarrow (E!m \rightarrow \varphi[m/x])$ is as stated. Let y be a variable that does not occur in the formula. Then $\psi \rightarrow (E!y \rightarrow \varphi[y/x])$ is a theorem. It follows that $\psi \rightarrow \forall y(E!y \rightarrow \varphi[y/x])$ is a theorem, and hence so is $\psi \rightarrow \forall x\varphi$. (b) follows from (a) and the fact that $\psi \rightarrow H\xi$ is a theorem of S iff $F\psi \rightarrow \xi$ is a theorem of S, and (c) from (a) and the fact that $\psi \rightarrow G\xi$ is a theorem of S iff $P\psi \rightarrow \xi$ is a theorem of S. For (d), suppose $\psi \rightarrow \text{At } n, (E!m \rightarrow \varphi[m/x])$ is as stated. Let y be a variable that does not occur in the formula. Then $\psi \rightarrow \text{At } n, (E!y \rightarrow \varphi[y/x])$ is a theorem. By (T3), it follows that (Sometimes, ψ) $\rightarrow$ At n, $(E!y \rightarrow \varphi[y/x])$ is a theorem, and hence so is (Sometimes, ψ) $\rightarrow \forall y\text{At } n, (E!y \rightarrow \varphi[y/x])$, and also therefore $\psi \rightarrow$ Always,$\forall y$At n, $(E!y \rightarrow \varphi[y/x])$. By (A24), it follows that $\psi \rightarrow \text{At } n, \forall y\varphi[y/x]$ is a theorem, and hence so is $\psi \rightarrow \text{At } n, \forall x\varphi$. □

Lemma 1 (Lindenbaum Lemma). Let S be a system that extends the neutral system, and let Δ be a set of £-formulas that is S-consistent. Then Δ can be extended to a set of £+-formulas which is nice relative to S.

Proof: Let Δ be as stated. Enumerate all the formulas of £+, and define a series Δ_0, Δ_1, … of sets of £+-formulas as follows:

1. $\Delta_0 = \Delta$
2. If φ_{k+1} is the (k+1)th £+-formula of our enumeration ($k \geq 0$), Δ_{n+1} is defined according to the following conditions:

   - If $\Delta_k \cup \{\varphi_{k+1}\}$ is inconsistent, then $\Delta_{k+1} = \Delta_k$
   - If $\Delta_k \cup \{\varphi_{k+1}\}$ is consistent and φ_{k+1} is neither of the form $\neg\forall x\varphi$, nor of the form $\neg H\forall x\varphi$, nor of the form $\neg G\forall x\varphi$, nor of the form $\neg\text{At } n, \forall x\varphi$, then $\Delta_{k+1} = \Delta_k \cup \{\varphi_{k+1}\}$

- If $\Delta_k \cup \{\varphi_{k+1}\}$ is consistent and φ_{k+1} is of the form $\neg\forall x\varphi$, then $\Delta_{k+1} = \Delta_k \cup \{\varphi_{k+1}\} \cup \{\neg(E!m \rightarrow \varphi[m/x])\}$, where m is the first constant in C+ not to appear in Δ_k or in φ_{k+1}
- If $\Delta_k \cup \{\varphi_{k+1}\}$ is consistent and φ_{k+1} is of the form $\neg H\forall x\varphi$, then $\Delta_{k+1} = \Delta_k \cup \{\varphi_{k+1}\} \cup \{\neg H(E!m \rightarrow \varphi[m/x])\}$, where m is the first constant in C+ not to appear in Δ_k or in φ_{k+1}
- If $\Delta_k \cup \{\varphi_{k+1}\}$ is consistent and φ_{k+1} is of the form $\neg G\forall x\varphi$, then $\Delta_{k+1} = \Delta_k \cup \{\varphi_{k+1}\} \cup \{\neg G(E!m \rightarrow \varphi[m/x])\}$, where m is the first constant in C+ not to appear in Δ_k or in φ_{k+1}
- If $\Delta_k \cup \{\varphi_{k+1}\}$ is consistent and φ_{k+1} is of the form $\neg$At n, $\forall x\varphi$, then $\Delta_{k+1} = \Delta_k \cup \{\varphi_{k+1}\} \cup \{\neg\text{At } n, (E!m \rightarrow \varphi[m/x])\}$, where m is the first constant in C+ not to appear in Δ_k or in φ_{k+1}

Define Δ^* as the union of all the Δ_ks with $k \geq 0$. It is clear from the construction that Δ^* extends Δ, is $\pounds$+-maximal, and has all the saturation properties involved in the definition of niceness. Δ^* is also consistent. This follows from the fact that for all $k \geq 0$, if Δ_k is consistent, then so is Δ_{k+1}. This latter point follows from facts (a) to (d) above. Δ^* is therefore nice. □

Let S be a system that extends the neutral system, and let Δ be a set of $\pounds$+-formulas that is nice relative to S. Where m is a constant of $\pounds$+, we let $\underline{m}$ be the set of all $\pounds$+-constants n such that $m = n \in \Delta$. We build a model $M^\Delta = \langle Ti^\Delta, Bef^\Delta, D^\Delta, Loc^\Delta, I^\Delta \rangle$ based on Δ by putting:

- $Ti^\Delta = \{\underline{m} : m$ an $\pounds$+-constant and $Tm \in \Delta\}$
- $\underline{m}\ Bef^\Delta\ \underline{n}$ iff $m \prec n \in \Delta$
- $D^\Delta(\underline{m}) = \{\underline{n} : n$ an $\pounds$+-constant and At m, $E!n \in \Delta\}$
- $Loc^\Delta(\underline{m}) = \{\underline{n} : n$ an $\pounds$+-constant and $n\ L\ m \in \Delta\}$
- $I^\Delta(m) = \underline{m}$, and $I^\Delta(\Phi, \underline{m}) = \{\langle \underline{m}_1, \ldots, \underline{m}_k\rangle :$ At m, $\Phi m_1 \ldots m_k \in \Delta\}$ where Φ is a k-place predicate distinct from =, T, $\prec$ and L

Lemma 2. M^Δ is a neutral model.

Proof: By (A28) Bef^Δ is a binary relation on Ti^Δ, and by (A8) and (A25) both D^Δ and Loc^Δ are functions defined on Ti. We verify the remaining conditions in turn.

- Ti^Δ has at least two elements.
  Using (A30), (A33) and (A34), one can show that At x, (PT $\vee$ FT) $\rightarrow$ Sometimes, $\exists y(Ty \,\&\, y \neq x)$ is a theorem of S. Using (A22), one can also show that (PT $\vee$ FT) $\rightarrow \exists x(Tx \,\&$ At x, (PT $\vee$ FT)) is a theorem of S, and hence, using (A14), that the same goes for $\exists x(Tx \,\&$ At x, (PT $\vee$ FT)). We infer that $\exists x(Tx \,\&$ Sometimes, $\exists y(Ty \,\&\, y \neq x))$ is a theorem of S, and therefore belongs to Δ. By $\exists$-saturation, then, for some constant m, $Tm \,\&$ Sometimes, $\exists y(Ty \,\&\, y \neq m)) \in \Delta$, and so both $Tm \in \Delta$ and Sometimes, $\exists y(Ty \,\&\, y \neq m) \in \Delta$. By Sometimes-$\exists$-saturation, it follows that for some constant n, Sometimes, $(Tn \,\&\, n \neq m) \in \Delta$, and so both $Tn \in \Delta$ and $n \neq m \in \Delta$. The result follows.

- Bef^{Δ} is irreflexive, transitive and total.
  This is due to (A30), (A31) and (A32).
- For all $u \in \mathrm{Ti}^{\Delta}$, $u \in \mathrm{D}^{\Delta}(u)$.
  This is due to axiom (A23).
- For all $u, v \in \mathrm{Ti}^{\Delta}$, $u \in \mathrm{Loc}^{\Delta}(v)$ iff $u = v$.
  This is due to axiom (A26).
- Function I^{Δ} takes each constant into an element of some $\mathrm{D}^{\Delta}(u)$, and each k-place predicate distinct from =, T, $\prec$ and L and each time u into a set of k-tuples of objects taken from the $\mathrm{D}^{\Delta}(u)$s.
  This follows from the definition of I^{Δ}. □

Suppose given an enumeration of all the constants of £+. Where φ is an £+-formula and r a variable-assignment, we let $[\varphi]^r$ be the result of replacing each occurrence of a free variable x in φ by the constant $m \in r(x)$ which is the first in our enumeration.

Lemma 3 (Truth Lemma). For all £+-formulas φ, times $\underline{n}$ in Ti^{Δ}, and assignments r: $\underline{n} \vDash_r \varphi$ iff At n, $[\varphi]^r \in \Delta$.

Proof: By induction on the complexity of the formulas. (For the sake of readability, we omit the superscript 'Δ' in the names of the elements of M^{Δ}.)

Atoms involving neither of =, T, $\prec$ and L. Consider for illustration an atomic formula Φmx where m is a constant. By the truth-clause for atomic formulas, $\underline{n} \vDash_r \Phi mx$ iff $\langle \mathrm{I}(m), r(x)\rangle \in \mathrm{I}(\Phi, \underline{n})$. Now $\mathrm{I}(m) = \underline{m}$ and $r(x) = \underline{m^*}$ for some constant m^*. We then have: $\langle \mathrm{I}(m), r(x)\rangle \in \mathrm{I}(\Phi, \underline{n})$ iff At n, $\Phi mm^* \in \Delta$. By the definition of $[\Phi mx]^r$, At n, $\Phi mm^* \in \Delta$ iff At n, $[\Phi mx]^r \in \Delta$.

Identity. Consider for illustration a formula $m = x$ where m is a constant. By the truth-clause for =, $\underline{n} \vDash_r m = x$ iff $\mathrm{I}(m) = r(x)$. Now $\mathrm{I}(m) = \underline{m}$ and $r(x) = \underline{m^*}$ for some constant m^*. We then have: $\mathrm{I}(m) = r(x)$ iff $m = m^* \in \Delta$. Since $m = m^* \to (\mathrm{T}n \to$ At n, $m = m^*)$ and (At n, $m = m^*) \to m = m^*$ are theorems of S, $m = m^* \in \Delta$ iff At n, $m = m^* \in \Delta$. By the definition of $[m = x]^r$, At n, $m = m^* \in \Delta$ iff At n, $[m = x]^r \in \Delta$.

Time. Consider for illustration a formula $\mathrm{T}m$ where m is a constant. By the truth-clause for T, $\underline{n} \vDash_r \mathrm{T}m$ iff $\mathrm{I}(m) \in \mathrm{Ti}$. Now $\mathrm{I}(m) = \underline{m}$. We then have: $\mathrm{I}(m) \in \mathrm{Ti}$ iff $\mathrm{T}m \in \Delta$. Since $\mathrm{T}m \to (\mathrm{T}n \to$ At n, $\mathrm{T}m)$ and (At n, $\mathrm{T}m) \to \mathrm{T}m$ are theorems of S, $\mathrm{T}m \in \Delta$ iff At n, $\mathrm{T}m \in \Delta$. By the definition of $[\mathrm{T}m]^r$, At n, $\mathrm{T}m \in \Delta$ iff At n, $[\mathrm{T}m]^r \in \Delta$.

Precedence. Just like identity, since $m \prec m^* \to (\mathrm{T}n \to$ At n, $m \prec m^*)$ and (At n, $m \prec m^*) \to m \prec m^*$ are theorems of S.

Location. Just like identity and precedence, since $m \,\mathrm{L}\, m^* \to (\mathrm{T}n \to$ At n, $m \,\mathrm{L}\, m^*)$ and (At n, $m \,\mathrm{L}\, m^*) \to m \,\mathrm{L}\, m^*$ are theorems of S.

Negation and conjunction. The inductive steps for negation and conjunction are secured thanks to axiom (A11) and the fact that At n, (φ & ψ) → (At n, φ & At n, ψ) and (At n, φ & At n, ψ) → At n, (φ & ψ) are theorems of S.

Quantification. In order to fix ideas, assume that the free variables in φ distinct from x are the distinct variables $x_1, \ldots, x_n$, choose $m_1 \in r(x_1), \ldots, m_n \in r(x_n)$, and let φ^* be $\varphi[m_1/x_1]\ldots[m_n/x_n]$. Then (i) $\underline{n} \vDash_r \forall x\varphi$ iff $\underline{n} \vDash_s \forall x\varphi^*$ for any assignment s, (ii) At n, $[\varphi^*]^s \in \Delta$ iff At n, $\varphi^*[m/x] \in \Delta$ for any assignment s such that $s(x) = \underline{m}$, and (iii) At n, $[\forall x\varphi]^r \in \Delta$ iff At n, $\forall x\varphi^* \in \Delta$.

(A) Suppose that $\underline{n} \vDash_r \forall x\varphi$. Then $\underline{n} \vDash_r \forall x\varphi^*$, and so by IH, At n, $[\varphi^*]^s \in \Delta$ for all assignments s differing from r at most on x and such that $s(x) \in \mathrm{D}(\underline{n})$. We then have: At n, $\varphi^*[m/x] \in \Delta$ for all £+-constants m such that At n, E!$m \in \Delta$. By At-∀-saturation, it follows that At n, $\forall x\varphi^* \in \Delta$, and so, At n, $[\forall x\varphi]^r \in \Delta$.

(B) Conversely, suppose that At n, $[\forall x\varphi]^r \in \Delta$. Then At n, $\forall x\varphi^* \in \Delta$. Let s be an assignment differing from r at most on x and such that $s(x) \in \mathrm{D}(\underline{n})$. Let m be in $s(x)$. Then At n, E!$m \in \Delta$ and so, At n, $\varphi^*[m/x] \in \Delta$. But then, At n, $[\varphi^*]^s \in \Delta$, and hence by IH, $\underline{n} \vDash_s \varphi^*$. It follows that $\underline{n} \vDash_r \forall x\varphi^*$, and hence that $\underline{n} \vDash_r \forall x\varphi$.

H. Let x be a variable distinct from n. Then by (A33), At n, $[\mathrm{H}\varphi]^r \in \Delta$ iff the following condition holds:

(Ha) Tn & Always, $\forall x(x \prec n \rightarrow$ At x, $[\varphi]^r) \in \Delta$

Suppose (Ha) holds. Let m be an £+-constant such that $m \prec n \in \Delta$. Then Sometimes, (E!m & $m \prec n$) $\in \Delta$, and so by the second conjunct of (Ha), Sometimes At m, $[\varphi]^r \in \Delta$. But then, At m, $[\varphi]^r \in \Delta$. Conversely, suppose that for all £+-constants m such that $m \prec n \in \Delta$, At m, $[\varphi]^r \in \Delta$. Then for all £+-constants m, Always, ($m \prec n \rightarrow$ At m, $[\varphi]^r) \in \Delta$. By Always-∀-saturation, it follows that Always, $\forall x(x \prec n \rightarrow$ At x, $[\varphi]^r)$ $\in \Delta$. Hence, the second conjunct of (Ha) holds. Hence, (Ha) is equivalent to:

(Hb) T$n \in \Delta$, and for all £+-constants m, if $m \prec n \in \Delta$, then At m, $[\varphi]^r \in \Delta$

By definition of Bef and IH, (Hb) is equivalent to: $\underline{n} \in$ Ti, and for all $\underline{m} \in$ Ti, if $\underline{m}$ Bef $\underline{n}$, then $\underline{m} \vDash_r \varphi^r$. This is equivalent to: $\underline{n} \vDash_r \mathrm{H}\varphi$.

G. Like the previous case, but appealing to (A34) instead of (A33).

At. Let m be a term of £+. Let m^* be m if m is a constant, and let m^* be the first constant belonging to $r(m)$ in our enumeration otherwise. By (A8), (T4) and (T1), At n, [At m, $\varphi]^r \in \Delta$ iff At m^*, $[\varphi]^r \in \Delta$. By IH, the latter is equivalent to: $\underline{m^*} \vDash_r \varphi$. Thanks to the definition of m^*, this is equivalent to: $\underline{n} \vDash_r$ At m, φ. □

Theorem 2. Let S be a system extending the neutral system, and let φ be an £-formula which is not a theorem of S. Then there is a nice set of £+-formulas Δ such that φ does not hold in M^{Δ}.

Proof: Let S and φ be as stated. Define the sentence φ* as follows: (i) if no variable occurs free in φ, then φ* = φ, (ii) otherwise, if the free variables in φ are $x_1, \ldots, x_n$ (with $i \neq j \Rightarrow x_i \neq x_j$), φ* is the result of replacing each x_i by an £-constant m_i not in φ (with $i \neq j \Rightarrow m_i \neq m_j$). Then φ* is not a theorem of S either, and, by (A22), nor is $\forall x(\mathrm{T}x \rightarrow \mathrm{At}\ x, \varphi^*)$. $\{\neg\forall x(\mathrm{T}x \rightarrow \mathrm{At}\ x, \varphi^*)\}$ is then S-consistent. Let Δ be a nice extension of $\{\neg\forall x(\mathrm{T}x \rightarrow \mathrm{At}\ x, \varphi^*)\}$ in language £+ (we know that there must be such a set by Lemma 1). Consider then the neutral model $\mathrm{M}^\Delta = \langle \mathrm{Ti}^\Delta, \mathrm{Bef}^\Delta, \mathrm{D}^\Delta, \mathrm{Loc}^\Delta, \mathrm{I}^\Delta \rangle$ based on Δ. By ∀-saturation, there is an $\underline{n} \in \mathrm{Ti}^\Delta$ such that At n, $\neg\varphi^* \in \Delta$. Let then r be any assignment whatever if φ has no free variables, and such that $m_i \in r(x_i)$, where the x_is and the m_is are as described above, otherwise. Then At n, $[\neg\varphi]^r \in \Delta$. By Lemma 3, it follows that $\underline{n} \vDash_r \neg\varphi$, and hence $\underline{n} \nvDash_r \varphi$. Hence, φ does not hold in M^Δ. □

Theorem 3 (Completeness of the neutral system). Every £-formula which holds in all neutral models is a theorem of the neutral system.

Proof: By Theorem 2 and Lemma 2. □

7. Completeness of the systems for GBT, presentism and permanentism

Given Theorem 2, in order to establish the completeness of the systems for GBT, presentism and permanentism relative to the associated semantics, it is enough to verify, for each system, that its characteristic axioms guarantee that for all nice sets of £ +-formulas Δ, the model M^Δ based on Δ is a model for the system.

GBT. We need to show that the following two conditions are satisfied:

- For all constants m and n, if $\underline{m}$ Bef^Δ $\underline{n}$, then $\mathrm{D}^\Delta(\underline{m}) \subseteq \mathrm{D}^\Delta(\underline{n})$
- For all constants m and n, if $\underline{n}$ Bef^Δ $\underline{m}$, then $\underline{m} \notin \mathrm{D}^\Delta(\underline{n})$

The first condition is secured by the fact that $m \prec n \rightarrow (\mathrm{At}\ m, \mathrm{E}!m^* \rightarrow \mathrm{At}\ n, \mathrm{E}!m^*)$ is a theorem of GBT. This in turn can be established using (P1) and (A34) (and other neutral principles). The second condition is secured by the fact that $n \prec m \rightarrow \neg\, \mathrm{At}\ n, \mathrm{E}!m$ is a theorem of GBT. This in turn can be established using (P2) and (A33) (and other neutral principles).

Presentism. We need to show that the following two conditions are satisfied:

- For all constants m and n, if $\underline{m}$ Bef^Δ $\underline{n}$, then $\underline{m} \notin \mathrm{D}^\Delta(\underline{n})$
- For all constants m and n, if $\underline{n}$ Bef^Δ $\underline{m}$, then $\underline{m} \notin \mathrm{D}^\Delta(\underline{n})$

We just saw that thanks to (P2) and (A33), the second condition is satisfied. The first condition is secured by the fact that $m \prec n \rightarrow \neg\, \mathrm{At}\ n, \mathrm{E}!m$ is a theorem of presentism. This in turn can be established using (P3) and (A34) (and other neutral principles).

Permanentism. We need to show that the following condition is satisfied:

- For all constants m and n such that $\underline{n} \in \mathrm{Ti}^\Delta$, $\underline{m} \in \mathrm{D}^\Delta(\underline{n})$

This is secured by the fact that $\mathrm{T}n \rightarrow \mathrm{At}\ n, \mathrm{E!}m$ is a theorem of permanentism. This in turn can be established using (PER′) (and neutral principles).

We can thus conclude:

Theorem 4 (Completeness of the non-neutral systems). Let S be any of the three non-neutral systems. Every $\mathcal{L}$-formula which holds in all the models for S is a theorem of S.

Appendix 2
Semantic Characterisation of the Relativistic Systems

The structure of this appendix is essentially the same as that of Appendix 1. Here there are not four but *eight* systems to take care of: the neutral system, the basic system for GBT and its pointy and bow-tie extensions, the basic system for presentism and its pointy and bow-tie extensions, and the system for permanentism.

1. The languages

Each system can be formulated in a language whose vocabulary is as specified in Appendix 1, section 1, save for the following differences:

- The 1-place predicate T for times is replaced by the predicate S for spacetime-points
- The 2-place predicate $\prec$ for temporal precedence is replaced by the 2-place predicate PREC for causal precedence
- The 2-place predicate L for location at a time is replaced by the 2-place predicate LOC for location at a spacetime-point
- The monadic temporal operators H and G are replaced by the monadic spatiotemporal operators ▲ and ▼, respectively, and a new monadic spatiotemporal operator ◀ is introduced
- The temporal prenective At is replaced by the spatiotemporal prenective @

The *terms* of such a language are defined as before, and its *formulas* are recursively defined as follows:

- If Φ is an n-place predicate and $m_1, \ldots, m_n$ are terms, then $\Phi m_1 \ldots m_n$ is a formula
- If φ is a formula, then so are $\neg\varphi$, ▲φ, ▼φ and ◀φ
- If φ and ψ are formulas, then so is $(\varphi \,\&\, \psi)$
- If φ is a formula and x a variable, then $\forall x\varphi$ is a formula
- If φ is a formula and m a term, then $@m\varphi$ is a formula

F. Correia, S. Rosenkranz, *Nothing To Come*, Synthese Library 395,
https://doi.org/10.1007/978-3-319-78704-6

The existence predicate E! is defined as before, and the following definitions are adopted:

- $\vartriangle\varphi \equiv_{df} \neg\blacktriangle\neg\varphi$
- $\triangledown\varphi \equiv_{df} \neg\blacktriangledown\neg\varphi$
- $\vartriangleleft\varphi \equiv_{df} \neg\blacktriangleleft\neg\varphi$
- $\bullet\varphi \equiv_{df} \blacktriangle\varphi \;\&\; \varphi \;\&\; \blacktriangleleft\varphi \;\&\; \blacktriangledown\varphi$
- $\circ\varphi \equiv_{df} \vartriangle\varphi \vee \varphi \vee \vartriangleleft\varphi \vee \triangledown\varphi$
- m SEP $n \equiv_{df} Sm \;\&\; Sn \;\&\; \neg(m = n \vee m$ PREC $n \vee n$ PREC $m)$

Up to section 6, we will suppose as given a fixed language £ of the sort just defined.

2. The neutral system

The axioms and rules of the neutral system are as follows (see Chap. 9).

Classical propositional axioms and rule:

As for the classical system.

Axioms and rule for quantification and identity:

As for the classical system, but with (A16) replaced by

($A16_R$) $\circ E!m$

Axioms for S:

($A9_R$) $Sx \rightarrow \bullet Sx$
($A23_R$) $Sx \rightarrow @xE!x$

Axioms for PREC:

($A28_R$) x PREC $y \rightarrow (Sx \;\&\; Sy)$
($A29_R$) x PREC $y \rightarrow \bullet(x$ PREC $y)$
($A30_R$) $\neg(x$ PREC $x)$
($A31_R$) $(x$ PREC $y \;\&\; y$ PREC $z) \rightarrow x$ PREC z

Axioms for LOC:

($A25_R$) x LOC $y \rightarrow Sy$
($A26_R$) $Sx \rightarrow (x$ LOC $y \leftrightarrow x = y)$
($A27_R$) x LOC $y \rightarrow \bullet(x$ LOC $y)$

Axioms and rules for ▲, ▼ and ◀:

($A1_R$) $\varphi \rightarrow \blacktriangle\triangledown\varphi$
($A2_R$) $\varphi \rightarrow \blacktriangledown\vartriangle\varphi$
(AR5) $\varphi \rightarrow \blacktriangleleft\vartriangleleft\varphi$

($A3_R$)	$\blacktriangle(\varphi \to \psi) \to (\blacktriangle\varphi \to \blacktriangle\psi)$
($A4_R$)	$\blacktriangledown(\varphi \to \psi) \to (\blacktriangledown\varphi \to \blacktriangledown\psi)$
(AR1)	$\blacktriangleleft(\varphi \to \psi) \to (\blacktriangleleft\varphi \to \blacktriangleleft\psi)$
($A5_R$)	$\triangledown\vartriangle\varphi \to (\vartriangle\varphi \vee \varphi \vee \triangleleft\varphi \vee \triangledown\varphi)$
($A6_R$)	$\vartriangle\triangledown\varphi \to (\vartriangle\varphi \vee \varphi \vee \triangleleft\varphi \vee \triangledown\varphi)$
(AR2)	$\triangleleft\triangleleft\varphi \to (\vartriangle\varphi \vee \varphi \vee \triangleleft\varphi \vee \triangledown\varphi)$
(AR3)	$(\triangleleft\triangledown\varphi \vee \triangledown\triangleleft\varphi) \to (\triangleleft\varphi \vee \triangledown\varphi)$
(AR4)	$(\triangleleft\vartriangle\varphi \vee \vartriangle\triangleleft\varphi) \to (\triangleleft\varphi \vee \vartriangle\varphi)$
($A7_R$)	$\vartriangle\vartriangle\varphi \to \vartriangle\varphi$
($A14_R$)	$\bigcirc\vartriangle\top$, for $\top$ any chosen tautology
($R1_R$)	$\varphi \;/\; \blacktriangle\varphi$
($R2_R$)	$\varphi \;/\; \blacktriangledown\varphi$
(RR)	$\varphi \;/\; \blacktriangleleft\varphi$

Axioms for @:

($A8_R$)	$@x\varphi \to Sx$
($A10_R$)	$@x(\varphi \to \psi) \to (@x\varphi \to @x\psi)$
($A11_R$)	$@x\neg\varphi \leftrightarrow (Sx \mathbin{\&} \neg @x\varphi)$
($A12_R$)	$\bullet\varphi \to (Sx \to @x\varphi)$
($A13_R$)	$@x\varphi \to \bullet @x\varphi$
($A22_R$)	$\forall x(Sx \to @x\varphi) \to \varphi$
($A33_R$)	$@x\blacktriangle\varphi \leftrightarrow Sx \mathbin{\&} \bullet\forall y(y \text{ PREC } x \to @y\varphi)$
($A34_R$)	$@x\blacktriangledown\varphi \leftrightarrow Sx \mathbin{\&} \bullet\forall y(x \text{ PREC } y \to @y\varphi)$
(AR6)	$@x\blacktriangleleft\varphi \leftrightarrow Sx \mathbin{\&} \bullet\forall y(y \text{ SEP } x \to @y\varphi)$
($A24_R$)	$@x\forall y\varphi \leftrightarrow \bullet\forall y @x(E!y \to \varphi)$

In ($A22_R$), x is not free in φ. In ($A33_R$), ($A34_R$) and (AR6), y is not free in φ. In ($A33_R$), ($A34_R$), ($A24_R$) and (AR6), x and y must be distinct variables.

3. The systems for GBT, presentism and permanentism

The basic system for GBT, GBT^{basic}, is defined by the addition of the following axioms to the neutral system:

($P1_R$)	$E!x \to \blacktriangledown E!x$
($P2_R$)	$Sx \to @x\blacktriangle\neg E!x$

GBT^{pointy} is GBT^{basic} plus axiom (PO), and $GBT^{bow\text{-}tie}$ is GBT^{basic} plus axiom (BO):

(PO)	$Sx \to @x\blacktriangleleft\neg E!x$
(BO)	$Sx \to @x\blacktriangleleft E!x$

The basic system for presentism, $PRES^{basic}$, is defined by the addition of the following axioms to the neutral system:

($P2_R$) $Sx \to @x\blacktriangle\neg E!x$
($P3_R$) $Sx \to @x\blacktriangledown\neg E!x$

$PRES^{pointy}$ is $PRES^{basic}$ plus axiom (PO), and $PRES^{bow\text{-}tie}$ is $PRES^{basic}$ plus axiom (BO). Finally, the system for permanentism results from the neutral system by adding axiom (PER′).

4. Semantics

The *neutral models* for language £ are the same kind of structures as the neutral models for the classical systems, namely tuples ⟨ Ti, Bef, D, Loc, I ⟩ satisfying certain conditions. (We shall use the same labels as in the classical case for the various elements of a neutral model, but of course, here the set Ti of such a structure is thought of as a set of spacetime-points rather than times, Bef is thought of as the relation of causal precedence between such points, and so on.) These conditions are almost the same as in the classical case, the differences boiling down to the following:

- The causal precedence relation of a neutral model is only required to be irreflexive and transitive (totality is not required)
- The set of spacetime-points of a neutral model is required to have at least two elements *standing in the causal precedence relation of that model* (the corresponding condition in classical models is automatically satisfied since in these models the temporal precedence relation is required to be total)
- The predicates which are interpreted by the interpretation function of a neutral model are those which are distinct from =, S, PREC and LOC

The models for GBT^{basic} are the neutral models that satisfy conditions (p1) and (p2). Let us use 'u Sep v' as short for '$u, v \in$ Ti, and neither $u = v$, nor u Bef v, nor v Bef u'. Adding the following condition defines the models for GBT^{pointy}:

(po) For all u, v such that u Sep v, $u \notin D(v)$
(At any time, no non-causally separated points exist)

Adding the following condition defines the models for $GBT^{bow\text{-}tie}$:

(bo) For all u, v such that u Sep v, $u \in D(v)$
(At any time, all the non-causally separated points exist)

The models for $PRES^{basic}$ are the neutral models that satisfy conditions (p2) and (p3). Adding condition (po) defines the models for $PRES^{pointy}$, and adding condition (bo) defines the models for $PRES^{bow\text{-}tie}$. Finally, the models for permanentism are the neutral models that satisfy condition (per′).

The truth-conditions for the formulas in a neutral model ⟨ Ti, Bef, D, Loc, I ⟩ are as follows:

[L1] For Φ a k-place predicate distinct from =, S, PREC and LOC:
$u \vDash_r \Phi m_1 \ldots m_k$ iff $\langle I_r(m_1), \ldots, I_r(m_k)\rangle \in I(\Phi, u)$

[L2] $u \models_r m = m^*$ iff $I_r(m)$ is $I_r(m^*)$
[L3] $u \models_r Sm$ iff $I_r(m) \in \mathrm{Ti}$
[L4] $u \models_r m$ PREC m^* iff $I_r(m)$ Bef $I_r(m^*)$
[L5] $u \models_r m$ LOC m^* iff $I_r(m) \in \mathrm{Loc}(I_r(m^*))$
[L6] $u \models_r \neg\varphi$ iff $u \not\models_r \varphi$
[L7] $u \models_r \varphi \,\&\, \psi$ iff both $u \models_r \varphi$ and $u \models_r \psi$
[L8] $u \models_r \forall x\varphi$ iff $u \models_s \varphi$ for all assignments s differing from r at most on x and such that $s(x) \in D(u)$.
[L9] $u \models_r \blacktriangle\varphi$ iff $v \models_r \varphi$ for all v such that v Bef u
[L10] $u \models_r \blacktriangledown\varphi$ iff $v \models_r \varphi$ for all v such that u Bef v
[L12] $u \models_r \blacktriangleleft\varphi$ iff $v \models_r \varphi$ for all v such that v Sep u
[L12] $u \models_r @m\varphi$ iff $I_r(m) \models_r \varphi$

A formula is said to *hold* in a model iff it is true at all spacetime-points of the model relative to all variable-assignments.

5. Soundness of the systems

Each system is sound with respect to its associated semantics:

Theorem 1 (Soundness of the eight systems). Let S be any of the systems defined in sections 2 and 3. Every theorem of S holds in all models for S.

The proof is left to the reader.

6. Completeness of the neutral system

The completeness of the relativistic neutral system is established exactly like the completeness of the classical neutral in section 6 of Appendix 1, and we will not bother here to give as many details as we gave there.

We consider a language £+, which is £ plus a countably infinite set C+ of new constants, and we assume as given a numbering of the constants in C+.

We keep the definitions of £+-maximality, S-consistency and ∀-saturation from Appendix 1. A set Δ of £+-formulas is said to be *nice* relative to a system S extending the neutral system iff:

- Δ is £+-maximal
- Δ is S-consistent
- Δ is ∀-saturated
- Δ is ▲-∀-saturated, i.e. the £+-formula $\blacktriangle\forall x\varphi$ belongs to Δ provided that $\blacktriangle(E!m \to \varphi[m/x])$ does for all £+-constants m
- Δ is ▼-∀-saturated, i.e. the £+-formula $\blacktriangledown\forall x\varphi$ belongs to Δ provided that $\blacktriangledown(E!m \to \varphi[m/x])$ does for all £+-constants m
- Δ is ◀-∀-saturated, i.e. the £+-formula $\blacktriangleleft\forall x\varphi$ belongs to Δ provided that $\blacktriangleleft(E!m \to \varphi[m/x])$ does for all £+-constants m
- Δ is @-∀-saturated, i.e. the £+-formula $@n\forall x\varphi$ belongs to Δ provided that $@n(E!m \to \varphi[m/x])$ does for all £+-constants m

Nice sets of $\pounds$+-formulas have the further properties of ●-∀-, ∃-, △-∃-, ▽-∃-, ◁-∃-, @-∃- and ○-∃-saturation (the second property is defined in Appendix 1; the way the other properties are to be defined should be obvious).

Let S be a system extending the neutral system. Then we have the following five facts, which will help establish Lemma 1 below:

(a) If $\psi \rightarrow (E!m \rightarrow \varphi[m/x])$ is a theorem of S, where m is a constant that appears neither in φ nor in ψ, then $\psi \rightarrow \forall x\varphi$ is also a theorem of S
(b) If $\psi \rightarrow \blacktriangle(E!m \rightarrow \varphi[m/x])$ is a theorem of S, where m is a constant that appears neither in φ nor in ψ, then $\psi \rightarrow \blacktriangle\forall x\varphi$ is also a theorem of S
(c) If $\psi \rightarrow \blacktriangledown(E!m \rightarrow \varphi[m/x])$ is a theorem of S, where m is a constant that appears neither in φ nor in ψ, then $\psi \rightarrow \blacktriangledown\forall x\varphi$ is also a theorem of S
(d) If $\psi \rightarrow \blacktriangleleft(E!m \rightarrow \varphi[m/x])$ is a theorem of S, where m is a constant that appears neither in φ nor in ψ, then $\psi \rightarrow \blacktriangleleft\forall x\varphi$ is also a theorem of S
(e) If $\psi \rightarrow @n(E!m \rightarrow \varphi[m/x])$ is a theorem of S, where m is a constant distinct from n that appears neither in φ nor in ψ, then $\psi \rightarrow @n\forall x\varphi$ is also a theorem of S

Proof: (a) See Appendix 1. (b) From (a) and the fact that $\psi \rightarrow \blacktriangle\xi$ is a theorem of S iff $\triangledown\psi \rightarrow \xi$ is a theorem of S. (c) From (a) and the fact that $\psi \rightarrow \blacktriangledown\xi$ is a theorem of S iff $\vartriangle\psi \rightarrow \xi$ is a theorem of S. (d) From (a) and the fact that $\psi \rightarrow \blacktriangleleft\xi$ is a theorem of S iff $\triangleleft\psi \rightarrow \xi$ is a theorem of S. (e) Similar to the proof of point (d) in Appendix 1, section 6. □

Lemma 1 (Lindenbaum Lemma). Let S be a system that extends the neutral system, and let Δ be a set of $\pounds$-formulas that is S-consistent. Then Δ can be extended to a set of $\pounds$+-formulas which is nice relative to S.

Proof: Like in Appendix 1, taking into consideration that in the relativistic languages, the three temporal operators H, G and At are replaced by ▲, ▼ and @, respectively, and that the relativistic languages have in addition the operator ◀. □

Let S be a system that extends the neutral system, and let Δ be a set of $\pounds$+-formulas that is nice relative to S. We define the model $M^{\Delta} = \langle Ti^{\Delta}, Bef^{\Delta}, D^{\Delta}, Loc^{\Delta}, I^{\Delta} \rangle$ based on Δ as we did in Appendix 1, *mutatis mutandis*.

Lemma 2. M^{Δ} is a neutral model.

Proof: Almost like in Appendix 1. The only significant differences concern the cardinality condition on Ti^{Δ} and the conditions on Bef^{Δ}:

- Ti^{Δ} has at least two elements standing in relation Bef.
  By ($A22_R$), $\bigcirc\vartriangle T \rightarrow \bigcirc\exists x(Sx\ \&\ @x\vartriangle T)$ is a theorem, and so by ($A14_R$), $\bigcirc\exists x(Sx\ \&\ @x\vartriangle T)$ is also a theorem. By ($A33_R$), ($A30_R$) and ($A8_R$), it follows that $\bigcirc\exists x(Sx\ \&\ \bigcirc\exists y(Sy\ \&\ y \neq x))$ is a theorem of S and therefore belongs to Δ. By ○-∃-satura-

tion, then, for some constant m, $\bigcirc(Sm \,\&\, \bigcirc\exists y(Sy \,\&\, y \neq m)) \in \Delta$, and so both $Sm \in \Delta$ and $\bigcirc\exists y(Sy \,\&\, y \neq m) \in \Delta$. By $\bigcirc$-$\exists$-saturation again, it follows that for some constant n, $\bigcirc(Sn \,\&\, n \neq m) \in \Delta$, and so both $Sn \in \Delta$ and $n \neq m \in \Delta$. The result follows.

- Bef^{Δ} is irreflexive and transitive.
  This is due to ($A30_R$) and ($A31_R$). □

Where φ is an $\pounds$+-formula and r a variable-assignment, we define $[\varphi]^r$ as in Appendix 1.

Lemma 3 (Truth Lemma). For all $\pounds$+-formulas φ, points $\underline{n}$ in Ti^{Δ}, and assignments r: $\underline{n} \vDash_r \varphi$ iff $@n[\varphi]^r \in \Delta$. □

Proof: Like in Appendix 1.

Theorem 2. Let S be a system extending the neutral system, and let φ be an $\pounds$-formula which is not a theorem of S. Then there is a nice set of $\pounds$+-formulas Δ such that φ does not hold in M^{Δ}.

Proof: See Appendix 1. □

Theorem 3 (Completeness of the neutral system). Every $\pounds$-formula which holds in all neutral models is a theorem of the neutral system.

Proof: From Theorem 2 and Lemma 2. □

7. Completeness of the systems for GBT, presentism and permanentism

As in Appendix 1, we exploit Theorem 2.

$\underline{\mathrm{GBT}^{\mathrm{basic}}}$. We need to show that the following two conditions are satisfied:

- For all constants m and n, if $\underline{m}$ Bef^{Δ} $\underline{n}$, then $\mathrm{D}^{\Delta}(\underline{m}) \subseteq \mathrm{D}^{\Delta}(\underline{n})$
- For all constants m and n, if $\underline{n}$ Bef^{Δ} $\underline{m}$, then $\underline{m} \notin \mathrm{D}^{\Delta}(\underline{n})$

The first condition is secured by the fact that m PREC $n \rightarrow (@m\mathrm{E}!m^* \rightarrow @n\mathrm{E}!m^*)$ is a theorem of $\mathrm{GBT}^{\mathrm{basic}}$, which can be established using ($P1_R$) and ($A34_R$) (and other neutral principles). The second condition is secured by the fact that n PREC $m \rightarrow \neg @n\mathrm{E}!m$ is a theorem of $\mathrm{GBT}^{\mathrm{basic}}$, which can be established using ($P2_R$) and ($A33_R$) (and other neutral principles).

$\underline{\mathrm{GBT}^{\mathrm{pointy}}}$. Given the previous facts about $\mathrm{GBT}^{\mathrm{basic}}$, we just need to show that the following condition is satisfied:

- For all constants m and n, if $\underline{n}$ Sep^{Δ} $\underline{m}$, then $\underline{m} \notin \mathrm{D}^{\Delta}(\underline{n})$

The condition is secured thanks to the fact that n SEP $m \rightarrow \neg@n\text{E!}m$ is a theorem of GBT^{pointy}, which can be established using (PO) and (AR6) (and other neutral principles).

GBT$^{\text{bow-tie}}$. Here we just need to show that the following condition is satisfied:

- For all constants m and n, if $\underline{n}$ Sep^{Δ} $\underline{m}$, then $\underline{m} \in \text{D}^{\Delta}(\underline{n})$

The condition is secured thanks to the fact that n SEP $m \rightarrow @n\text{E!}m$ is a theorem of GBT$^{\text{bow-tie}}$, which can be established using (BO) and (AR6) (and other neutral principles).

PRES^{basic}. We need to show that the following two conditions are satisfied:

- For all constants m and n, if $\underline{m}$ Bef^{Δ} $\underline{n}$, then $\underline{m} \notin \text{D}^{\Delta}(\underline{n})$
- For all constants m and n, if $\underline{n}$ Bef^{Δ} $\underline{m}$, then $\underline{m} \notin \text{D}^{\Delta}(\underline{n})$

We saw that thanks to (P2_{R}) and (A33_{R}), the second condition is satisfied. The first condition is secured by the fact that m PREC $n \rightarrow \neg@n\text{E!}m$ is a theorem of PRES^{basic}, which can be established using (P3_{R}) and (A34_{R}) (and other neutral principles).

PRES^{pointy}. As we saw, the presence of axiom (PO) guarantees that the following condition is satisfied:

- For all constants m and n, if $\underline{n}$ Sep^{Δ} $\underline{m}$, then $\underline{m} \notin \text{D}^{\Delta}(\underline{n})$

PRES$^{\text{bow-tie}}$. As we saw, the presence of axiom (BO) guarantees that the following condition is satisfied:

- For all constants m and n, if $\underline{n}$ Sep^{Δ} $\underline{m}$, then $\underline{m} \in \text{D}^{\Delta}(\underline{n})$

Permanentism. We need to show that the following condition is satisfied:

- For all constants m and n such that $\underline{n} \in \text{Ti}^{\Delta}$, $\underline{m} \in \text{D}^{\Delta}(\underline{n})$

This is secured by the fact that $\text{T}n \rightarrow @n\text{E!}m$ is a theorem of permanentism, which can be established using (PER′) (and neutral principles).

We can thus conclude:

Theorem 4 (Completeness of the non-neutral systems). Let S be any of the seven non-neutral systems. Every £-formula which holds in all the models for S is a theorem of S.

Appendix 3
Recovering the Classical Systems

Here we show that given the two principles about the causal structure of spacetime characteristic of prerelativistic physics introduced in Sect. 9.5, namely

(PR1) x PREC y & x SIM $z \rightarrow z$ PREC y

and

(PR2) $@x\varphi$ & x SIM $y \rightarrow @y\varphi$,

the classical systems defined in Appendix 1 'follow from' their relativistic counterparts. More precisely, we define a natural translation function τ from the classical language as described in Appendix 1 into a slightly enriched version of the relativistic language as described in Appendix 2, and we establish the following facts:

> If formula φ is a theorem of the classical neutral system/classical GBT/classical presentism/classical permanentism, then its translation $(\varphi)^\tau$ follows from (PR1) and (PR2) in the relativistic neutral system/basic relativistic GBT/basic relativistic presentism/relativistic permanentism.

The enriched relativistic language is defined by adding to the original relativistic language a single binary predicate ε for class membership, and a denumerably infinite stock of variables distinct from the variables of the original language. These variables are called *new*, and we assume that they are numbered. Define the predicate C for classes as follows:

- $Cm \equiv_{df} \bigcirc\exists x(x\ \varepsilon\ m)$
  (with x the first new variable distinct from m)

We adopt the following axioms for classes and class membership:

(AX1)	$x\ \varepsilon\ y \rightarrow Sx$	Restriction
(AX2)	$x\ \varepsilon\ y \rightarrow \bullet(x\ \varepsilon\ y)$	Eternality

F. Correia, S. Rosenkranz, *Nothing To Come*, Synthese Library 395,
https://doi.org/10.1007/978-3-319-78704-6

(AX3)	$Cx \mathbin{\&} Cy \mathbin{\&} \bullet\forall z(z \mathrel{\varepsilon} x \leftrightarrow z \mathrel{\varepsilon} y) \rightarrow x = y$	Identity conditions
(AX4)	$Cx \rightarrow (x \text{ LOC } y \leftrightarrow y \mathrel{\varepsilon} x)$	Location conditions
(AX5)	$Cx \rightarrow (@y\mathrm{E}!x \leftrightarrow \bullet\forall z(z \mathrel{\varepsilon} x \rightarrow @y\mathrm{E}!z))$	Existence conditions

(AX1) reflects the fact that we are only interested in classes of spacetime-points. The other axioms speak for themselves.

Define the predicate SIM for simultaneity between spacetime-points as follows:

- $x \text{ SIM } y \equiv_{\mathrm{df}} Sx \mathbin{\&} Sy \mathbin{\&} \neg\, x \text{ PREC } y \mathbin{\&} \neg\, y \text{ PREC } x$

We adopt a further axiom involving ε, which says that at any spacetime-point, there exists a class whose members are the spacetime-points that exist at, and are simultaneous with, that spacetime point:

(AX6) $Sx \rightarrow @x\exists y\bullet\forall z(z \mathrel{\varepsilon} y \leftrightarrow z \text{ SIM } x \mathbin{\&} @x\mathrm{E}!z))$ Restricted comprehension

This axiom follows from more general comprehension principles, but (AX6) is enough for our purposes.

We finally adopt the following definitions:

- $SCm \equiv_{\mathrm{df}} Cm \mathbin{\&} \bullet\forall x\bullet\forall y(x \mathrel{\varepsilon} m \mathbin{\&} y \mathrel{\varepsilon} m \rightarrow x \text{ SIM } y)$
  (with x and y respectively the first and second new variable distinct from m)
- $m \text{ SIMEM } n \equiv_{\mathrm{df}} \bigcirc\exists x(m \text{ SIM } x \mathbin{\&} x \mathrel{\varepsilon} n)$
  (with x the first new variable distinct from both m and n)
- $T^*m \equiv_{\mathrm{df}} SCm \mathbin{\&} \bullet\forall x(x \text{ SIMEM } m \rightarrow x \mathrel{\varepsilon} m)$
  (with x the first new variable distinct from m)
- $m \prec^* n \equiv_{\mathrm{df}} T^*m \mathbin{\&} T^*n \mathbin{\&} \bullet\forall x\bullet\forall y(x \mathrel{\varepsilon} m \mathbin{\&} y \mathrel{\varepsilon} n \rightarrow x \text{ PREC } y)$
  (with x and y respectively the first and second new variable distinct from both m and n)
- $m\, L^*\, n \equiv_{\mathrm{df}} T^*n \mathbin{\&} \bigcirc\exists x(x \mathrel{\varepsilon} n \mathbin{\&} m \text{ LOC } x)$
  (with x the first new variable distinct from both m and n)
- $At^*\, m, \varphi \equiv_{\mathrm{df}} T^*m \mathbin{\&} \bullet\forall x(x \mathrel{\varepsilon} m \rightarrow @x\varphi)$
  (with x the first new variable distinct from m and not free in φ)

'SC' is mnemonic for 'simultaneity class', and 'SIMEM' for 'is simultaneous with a member of'. The starred expressions will be used to translate their unstarred mates.

The translation function τ is defined via the following recursive specification:

(a) If φ is an atomic formula not containing T, $\prec$ or L, $(\varphi)^\tau$ is φ
(b) $(Tm)^\tau$ is T^*m
(c) $(m \prec n)^\tau$ is $m \prec^* n$
(d) $(m\, L\, n)^\tau$ is $m\, L^*\, n$
(e) $(\neg\varphi)^\tau$ is $\neg(\varphi)^\tau$
(f) $(\varphi \mathbin{\&} \psi)^\tau$ is $(\varphi)^\tau \mathbin{\&} (\psi)^\tau$
(g) $(H\varphi)^\tau$ is $\blacktriangle(\varphi)^\tau$
(h) $(G\varphi)^\tau$ is $\blacktriangledown(\varphi)^\tau$
(i) $(\forall x\varphi)^\tau$ is $\forall x(\varphi)^\tau$
(j) $(At\, m, \varphi)^\tau$ is $At^*\, m, (\varphi)^\tau$

Let us then first establish the following:

Fact 1. If formula φ is a theorem of the classical neutral system, then its translation $(\varphi)^{\tau}$ follows from (PR1) and (PR2) in the relativistic neutral system.

Proof: What needs to be done is (i) show that the translation of every axiom of the classical neutral system is derivable in the relativistic neutral system in the presence of (PR1) and (PR2), (ii) show that the translation of each classical rule is a rule that preserves theoremhood in the relativistic system. Point (ii) is immediate. For point (i), we give indications below about how to proceed for each axiom that deserves some attention. We will appeal to the fact that given (PR1) and (PR2), the following are derivable in the relativistic neutral system:

(TR1) $\lhd\varphi \rightarrow \varphi$
(TR3) $\mathrm{S}x \rightarrow @x\exists y(\mathrm{T}^{*}y \;\&\; x \,\varepsilon\, y)$
(A30#) $\neg(x \prec^{*} x)$
(A31#) $(x \prec^{*} y \;\&\; y \prec^{*} z) \rightarrow x \prec^{*} z$
(A32#) $(\mathrm{T}^{*}x \;\&\; \mathrm{T}^{*}y) \rightarrow (x \prec^{*} y \vee x = y \vee y \prec^{*} x)$

We established the derivability of (TR1), (A30#), (A31#) and (A32#) in Sect. 9.5. (TR3) can be derived using the comprehension axiom (AX6), (PR2) and the transitivity of SIM, which latter, as we saw in Sect. 9.5, follows from (PR1). Let us then run through the selected choice of axioms:

(A19) $\forall x\varphi \;\&\; \mathrm{E}!m \rightarrow \varphi[m/x]$
By the definition of τ, $(\forall x\varphi \;\&\; \mathrm{E}!m \rightarrow \varphi[m/x])^{\tau}$ is identical to $\forall x(\varphi)^{\tau} \;\&\; \mathrm{E}!m \rightarrow (\varphi[m/x])^{\tau}$. This formula is a theorem of relativistic neutral system, since provided that $\varphi[m/x]$ is defined, so is $(\varphi)^{\tau}[m/x]$, and the latter is logically equivalent to $(\varphi[m/x])^{\tau}$. This latter fact can be proved by induction on the complexity of φ.

- If φ is atomic and does not contain T, $\prec$ or L, $(\varphi)^{\tau}$ is φ and so the result is immediate. If φ is atomic and does contain T, $\prec$ or L, the result is not immediate. The readercan verify by herself that the fact that T*, $\prec^{*}$ and L* are defined using new variables, rather than variables of the original language, is crucial in this step of the proof.
- For the induction step, we need to run through the various forms φ can take: $\neg\psi$, $\psi \;\&\; \xi$, $\mathrm{H}\psi$, $\mathrm{G}\psi$, $\forall x\psi$ and At m, ψ. The first five cases are straightforward. Take the third case for instance. Suppose $(\mathrm{H}\psi)[m/x]$ is defined. Then so is $\psi[m/x]$. By induction hypothesis, $(\psi)^{\tau}[m/x]$ is then also defined, and hence so is $(\mathrm{H}\psi)^{\tau}[m/x]$. By induction hypothesis, $(\psi)^{\tau}[m/x]$ is logically equivalent to $(\psi[m/x])^{\tau}$. Hence, $\blacktriangle(\psi)^{\tau}[m/x]$ is logically equivalent to $\blacktriangle(\psi[m/x])^{\tau}$. But then by the definition of τ, $(\mathrm{H}\psi)^{\tau}[m/x]$ is logically equivalent to $((\mathrm{H}\psi)[m/x])^{\tau}$. We thus have the expected result. The last case to consider is less straightforward. As the reader can verify, the fact that At* is defined using new variables is crucial here.

(A16) Sometimes, E!m
(Sometimes, E!m)$^\tau$ is equivalent to E!m ∨ ▿E!m ∨ ▵E!m. This latter formula can be derived using ($A16_R$) and (TR1).

(A21) $x = y \rightarrow (\varphi \rightarrow \varphi[y//x])$
The proof is similar to the proof for (A19).

(A9) Tx → Always, Tx
T*x → ▲T*x and T*x → ▼T*x are derivable in the relativistic neutral system, thanks to the fact that T*x is a conjunction of formulas starting with ● or ○.

(A23) Tx → At x, E!x
We need to derive T*x → ●∀y(y ε x → @yE!x), where y is the first new variable distinct from x. This can be done using (AX5), (PR2) and ($A23_R$).

(A28) $x \prec y \rightarrow$ (Tx & Ty)
Immediate by the definition of ≺*

(A29) $x \prec y \rightarrow$ Always, ($x \prec y$)
$x \prec^* y$ → ▲($x \prec^* y$) and $x \prec^* y$ → ▼($x \prec^* y$) are derivable in the relativistic neutral system, thanks to the fact that $x \prec^* y$ is a conjunction of formulas starting with ● or ○.

(A30) $\neg(x \prec x)$
See (A30#).

(A31) $(x \prec y \;\&\; y \prec z) \rightarrow x \prec z$
See (A31#).

(A32) (Tx & Ty) → ($x \prec y \vee x = y \vee y \prec x$)
See (A32#).

(A25) x L y → Ty
Immediate by the definition of L*

(A26) Tx → (x L y ↔ $x = y$)
T*x → (x L* y → $x = y$) can be established using (AX3) and (AX4). T*x → ($x = y$ → x L* y) can be established using (AX4).

(A27) x L y → Always, x L y
x L* y → ▲(x L* y) and x L* y → ▼(x L* y) are derivable in the relativistic neutral system, thanks to the fact that x L* y is a conjunction of formulas starting with ● or ○.

(A5) $FP\varphi \rightarrow (P\varphi \vee \varphi \vee F\varphi)$
Use ($A5_R$) and (TR1).

(A6) $PF\varphi \rightarrow (P\varphi \vee \varphi \vee F\varphi)$
Use ($A6_R$) and (TR1).

(A14) $P\top \vee F\top$, for ⊤ any chosen tautology
Use ($A14_R$), ($A5_R$), ($A7_R$) and (TR1).

(A8) (At x, φ) → Tx
Immediate by the definition of At*.

(A10) At x, (φ → ψ) → (At x, φ → At x, ψ)
Use ($A10_R$).

(A11) At x, ¬φ ↔ (Tx & ¬At x, φ)

Use ($A11_R$) left-to-right to derive At* x, $\neg(\varphi)^\tau \rightarrow \neg$At* x, $(\varphi)^\tau$. Use ($A11_R$) right-to-left and (PR2) to derive (Tx & $\neg$At* x, $(\varphi)^\tau$) $\rightarrow$ At* x, $\neg(\varphi)^\tau$.

(A12) (Always, φ) $\rightarrow$ (Tx $\rightarrow$ At x, φ)
Given (TR1), (Always, $\varphi)^\tau \rightarrow \bullet(\varphi)^\tau$ can be derived. The result follows from the definition of At*.

(A13) (At x, φ) $\rightarrow$ Always, At x, φ
At* x, $(\varphi)^\tau \rightarrow \blacktriangle$At* x, $(\varphi)^\tau$ and At* x, $(\varphi)^\tau \rightarrow \blacktriangledown$At* x, $(\varphi)^\tau$ are derivable in the relativistic neutral system, thanks to the fact that At* x, $(\varphi)^\tau$ is a conjunction of formulas starting with $\bullet$ or $\circ$.

(A22) $\forall x$(Tx $\rightarrow$ At x, φ) $\rightarrow$ φ
Suppose @$z\forall x$(T*x $\rightarrow$ At* x, $(\varphi)^\tau$) with x not free in φ (and hence in $(\varphi)^\tau$) and z a variable that does not appear in $\forall x$(T*x $\rightarrow$ At* x, $(\varphi)^\tau$). By (TR4) we then have @$z\exists y$(T*y & z ε y & At* y, $(\varphi)^\tau$), and so @$z\exists y$(T*y & z ε y & $\bullet\forall z'$(z' ε y $\rightarrow$ @$z'(\varphi)^\tau$)). We infer @$z(\varphi)^\tau$. Thus, we have shown @$z\forall x$(T*x $\rightarrow$ At* x, $(\varphi)^\tau$) $\rightarrow$ @$z(\varphi)^\tau$. We infer Sz $\rightarrow$ @z($\forall x$(T*x $\rightarrow$ At* x, $(\varphi)^\tau$) $\rightarrow$ $(\varphi)^\tau$). Using ($A22_R$) we infer $\forall x$(T*x $\rightarrow$ At* x, $(\varphi)^\tau$) $\rightarrow$ $(\varphi)^\tau$.

(A33) At x, Hφ $\leftrightarrow$ Tx & Always, $\forall y$($y \prec x$ $\rightarrow$ At y, φ)
Given (TR1), it is enough to derive $\bullet\forall z$(z ε x $\rightarrow$ @$z\blacktriangle(\varphi)^\tau$) $\leftrightarrow$ $\bullet\forall y$($y \prec^* x$ $\rightarrow$ $\bullet\forall z$(z ε y $\rightarrow$ @$z(\varphi)^\tau$)). The left-to-right conditional can be derived using ($A33_R$) left-to-right. The right-to-left conditional can be derived using ($A33_R$) right-to-left, (TR4) and (PR1).

(A34) At x, Gφ $\leftrightarrow$ Tx & Always, $\forall y$($x \prec y$ $\rightarrow$ At y, φ)
Idem, but with ($A34_R$) replacing ($A33_R$).

(A24) At x, $\forall y\varphi$ $\leftrightarrow$ Always, $\forall y$ At x, (E!y $\rightarrow$ φ)
Use ($A24_R$). □

Let us now move on to establish the remaining facts on our list.

Fact 2. If formula φ is a theorem of classical GBT, then its translation $(\varphi)^\tau$ follows from (PR1) and (PR2) in the basic system for relativistic GBT.

Proof: We just need to establish that the characteristic axioms of classical GBT, namely

(P1) E!x $\rightarrow$ GE!x

and

(P2) Tx $\rightarrow$ At x, H$\neg$E!x

translate into formulas which are derivable in the basic system for relativistic GBT given (PR1) and (PR2). The point about (P1) is immediate since (P1)'s translation is ($P1_R$). For (P2), use the fact that $x \varepsilon y \rightarrow (Ey \rightarrow Ex)$ is derivable, which in turn can be established using (AX5). Notice that neither (PR1) nor (PR2) has been used here. □

Fact 3. If formula φ is a theorem of classical presentism, then its translation $(\varphi)^\tau$ follows from (PR1) and (PR2) in the basic system for relativistic presentism.

Proof: We just need to establish that the characteristic axioms of classical presentism, namely

(P2) $Tx \rightarrow \text{At } x, H\neg E!x$

and

(P3) $Tx \rightarrow \text{At } x, G\neg E!x$

translate into formulas which are derivable in the basic system for relativistic presentism given (PR1) and (PR2). For the point about (P2), see the case of GBT above. For (P3), use ($P3_R$) and the fact that $x \varepsilon y \rightarrow (Ey \rightarrow Ex)$ is derivable. Here again, neither (PR1) nor (PR2) has been used. □

Fact 4. If formula φ is a theorem of classical permanentism, then its translation $(\varphi)^\tau$ follows from (PR1) and (PR2) in the system for relativistic permanentism.

Proof: We just need to establish that the characteristic axiom of classical permanentism, namely

(PER′) $E!x$

translate into a formula which is derivable in the system for relativistic permanentism (PR1) and (PR2). This is immediate. Once again, the presence of (PR1) nor (PR2) is immaterial. □

Bibliography

Baia, A. (2012). Presentism and the grounding of truth. *Philosophical Studies, 159*, 341–356.

Barnes, E., & Cameron, R. (2009). The open future: bivalence, determinism and ontology. *Philosophical Studies, 146*, 291–309.

Belnap, N., Perloff, M., & Xu, M. (2001). *Facing the future. Agents and choices in our indeterminist world.* Oxford: Oxford University Press.

Bigelow, J. (1996). Presentism and properties. *Philosophical Perspectives, 10*, 35–52.

Blake, R. M. (1925). On Mr. Broad's theory of time. *Mind, 34*, 418–435.

Bourne, C. (2002). When am I? A tense time for some tense theorists? *Australasian Journal of Philosophy, 80*, 359–371.

Bourne, C. (2006). *A future for presentism*. Oxford: Oxford University Press.

Braddon-Mitchell, D. (2004). How do we know that it is now now? *Analysis, 64*, 199–203.

Braddon-Mitchell, D. (2013). Fighting the zombie of the growing salami. *Oxford Studies in Metaphysics, 8*, 351–361.

Broad, C. D. (1923). *Scientific thought*. London: Routledge.

Broad, C. D. (1937). The philosophical implications of foreknowledge, *Proceedings of the Aristotelian Society. Supplementary, 16*, 177–209.

Broad, C. D. (1938). *Examination of McTaggart's philosophy* (Vol. 2). Cambridge: Cambridge University Press.

Broad, C. D. (1959). A reply to my critics. In P. A. Schilpp (Ed.), *The philosophy of C. D. Broad* (pp. 711–830). New York: Tudor Publishing Company.

Button, T. (2006). There's no time like the present. *Analysis, 66*, 130–135.

Button, T. (2007). Every now and then, no-futurism faces no sceptical problems. *Analysis, 67*, 325–332.

Cameron, R. (2015). *The moving spotlight. An essay on time and ontology*. Oxford: Oxford University Press.

Correia, F., & Iacona, A. (2013). *Around the tree. Semantic and metaphysical issues concerning branching and the open future, Synthese Library 361*. Heidelberg: Springer.

Correia, F., & Rosenkranz, S. (2011). *As time goes by. Eternal facts in an ageing universe.* Paderborn: Mentis.

Correia, F., & Rosenkranz, S. (2012). Eternal facts in an ageing universe. *Australasian Journal of Philosophy, 90*, 307–320.

Correia, F., & Rosenkranz, S. (2013). Living on the brink, or welcome back, growing block! *Oxford Studies in Metaphysics, 8*, 333–350.

Correia, F., & Rosenkranz, S. (2015a). Return of the living dead. Reply to Braddon-Mitchell. *Oxford Studies in Metaphysics, 9*, 376–389.

Correia, F., & Rosenkranz, S. (2015b). Presentism without presentness. *Thought: A Journal of Philosophy, 4*, 19–27.

F. Correia, S. Rosenkranz, *Nothing To Come*, Synthese Library 395,
https://doi.org/10.1007/978-3-319-78704-6

Correia, F., & Schnieder, B. (2012). *Metaphysical grounding. Understanding the structure of reality*. Cambridge: Cambridge University Press.
Crisp, T. M. (2004). On presentism and triviality. *Oxford Studies in Metaphysics, 1*, 15–20.
Deasy, D. (2015). The moving spotlight theory. *Philosophical Studies, 172*, 2073–2089.
Dorr, C., & Goodman, J. (forthcoming). Diamonds are forever. *Noûs*.
Dummett, M. (2004). *Truth and the past*. New York: Columbia University Press.
Dummett, M. (2006). *Thought and reality*. Oxford: Oxford University Press.
Evans, G. (1985). Does tense logic rest on a mistake? In G. Evans (Ed.), *Collected papers* (pp. 343–363). Oxford: Oxford University Press.
Fine, K. (2005). *Modality and tense. Philosophical papers*. Oxford: Oxford University Press.
Forrest, P. (2004). The real but dead past: A reply to Braddon-Mitchell. *Analysis, 64*, 358–362.
Forrest, P. (2006). Uniform grounding of truth and the growing block theory: A reply to Heathwood. *Analysis, 66*, 161–163.
Gallois, A. (2004). Comments on Ted Sider: Four-dimensionalism. *Philosophy and Phenomenological Research, 68*, 648–657.
Hawking, S. W., & Ellis, G. F. R. (1973). *The large scale structure of space-time*. Cambridge: Cambridge University Press.
Heathwood, C. (2005). The real price of the dead past: A reply to Forrest and to Braddon-Mitchell. *Analysis, 65*, 249–251.
Hinchliff, M. (2000). A defense of presentism in a relativistic setting. *Philosophy of Science, 67*, 575–586.
Kamp, H. (1971). Formal properties of 'now'. *Theoria, 37*, 227–274.
Kierland, B., & Monton, B. (2007). Presentism and the objection from being-supervenience. *Australasian Journal of Philosophy, 85*, 485–497.
King, J. C. (2003). Tense, modality, and semantic values. *Philosophical Perspectives, 17*, 195–246.
Lewis, D. K. (1986). *On the plurality of worlds*. Oxford: Blackwell.
MacFarlane, J. (2003). Future contingents and relative truth. *The Philosophical Quarterly, 53*, 321–336.
Markosian, N. (1995). The open past. *Philosophical Studies, 79*, 95–105.
McTaggart, J. M. E. (1908). The unreality of time. *Mind, 17*, 457–474.
McTaggart, J. M. E. (1927). *The nature of existence, Vol. 2*. Cambridge: Cambridge University Press.
Mellor, H. (1998). *Real time II*. London: Routledge.
Merricks, T. (2006). Goodbye growing block. *Oxford Studies in Metaphysics, 2*, 103–110.
Merricks, T. (2007). *Truth and ontology*. Oxford: Oxford University Press.
Meyer, U. (2013a). *The nature of time*. Oxford: Oxford University Press.
Meyer, U. (2013b). The triviality of presentism. In R. Ciuni, K. Miller, & G. Torrengo (Eds.), *New papers on the present* (pp. 67–88). Munich: Philosophia Verlag.
Miller, K. (2013). The growing block, presentism and eternalism. In A. Bardon & H. Dyke (Eds.), *A companion to the philosophy of time* (pp. 345–365). Oxford: Blackwell.
Minguzzi, E., & Sánchez, M. (2008). The causal hierarchy of spacetimes. In D. V. Alekseevsky & H. Baum (Eds.), *Recent developments in pseudo-Riemannian geometry, ESI Lectures in Mathematics and Physics* (pp. 299–358). Zürich: EMS Publishing House.
Minkowski, H. (1952). Space and time. In H. A. Lorentz, A. Einstein, H. Minkowski, & H. Weyl (Eds.), *The principle of relativity: A collection of original memoirs on the special and general theory of relativity* (pp. 75–91). New York: Dover.
Moore, G. E. (1939). Proof of an external world. *Proceedings of the British Academy, 25*, 273–300.
Mulligan, K. (2011). Meaning something and meanings. *Grazer Philosophische Studien, 82*, 255–284.
Oaklander, L. N. (2004). *The ontology of time*. Amherst/NY: Prometheus Books.
Partee, B. H. (1973). Some structural analogies between tenses and pronouns in English. *The Journal of Philosophy, 70*, 601–609.

Prior, A. N. (1953). Three-valued logic and future contingents. *The Philosophical Quarterly, 3*, 317–326.
Prior, A. N. (1958). Time after time. *Mind, 67*, 244–246.
Prior, A. N. (1959). Thank goodness that's over. *Philosophy, 34*, 12–17.
Prior, A. N. (1967). *Past, present and future*. Oxford: Clarendon Press.
Prior, A. N. (1970). The notion of the present. *Studium Generale, 23*, 245–248.
Prior, A. N. (2003). *Papers on time and tense, new edition*. Oxford: Oxford University Press.
Putnam, H. (1962). The analytic and the synthetic. *Minnesota Studies in the Philosophy of Science, 3*, 358–397.
Putnam, H. (1967). Time and physical geometry. *The Journal of Philosophy, 64*, 240–247.
Reichenbach, H. (1947). *Elements of symbolic logic*. New York: Macmillan.
Rietdijk, C. W. (1966). A rigorous proof of determinism from the special theory of relativity. *Philosophy of Science, 33*, 341–344.
Rosenkranz, S. (2012). In defence of Ockhamism, *Philosophia – Philosophical Quarterly of Israel* 40, 617–631, reprinted as In defense of Ockhamism. In J. M. Fischer and P. Todd (Eds.), *Freedom, fatalism, and foreknowledge*, New York 2015: Oxford University Press, 343–360.
Rosenkranz, S. (2013). Determinism, the open future and branching time. In *Correia and Iacona 2013* (pp. 47–72).
Ross, W. D., et al. (1908). *The works of Aristotle, volume 8: Metaphysica*. Oxford: Clarendon Press.
Russell, B. (1912). *The problems of philosophy*. London: Williams and Norgate.
Savitt, S. F. (2000). There's no time like the present (in Minkowski Spacetime). *Philosophy of Science, 67* (Supplement), 563–574.
Shoemaker, S. (1969). Time without change. *The Journal of Philosophy, 66*, 363–381.
Sider, T. (2001). *Four-dimensionalism*. Oxford: Oxford University Press.
Sider, T. (2011). *Writing the book of the world*. Oxford: Oxford University Press.
Smart, J. J. C. (1949). The river of time. *Mind, 58*, 483–494.
Stein, H. (1968). On Einstein-Minkowski space-time. *The Journal of Philosophy, 65*, 5–23.
Stein, H. (1970). A note on time and relativity theory. *The Journal of Philosophy, 67*, 289–294.
Stein, H. (1991). On relativity theory and openness of the future. *Philosophy of Science, 58*, 147–167.
Tallant, J. (2007). There have been, are (now), and will be lots of times like the present in the hybrid view of time. *Analysis, 67*, 83–86.
Tallant, J. (2011). There's no future in no-futurism. *Erkenntnis, 74*, 37–52.
Tallant, J., & Ingram, D. (2015). Nefarious presentism. *The Philosophical Quarterly, 65*, 355–371.
Thomason, R. H. (1970). Indeterminist time and truth-value gaps. *Theoria, 36*, 264–281.
Tooley, M. (1997). *Time, tense and causation*. Oxford: Oxford University Press.
Weingard, R. (1972). Relativity and the reality of past and future events. *British Journal for the Philosophy of Science, 23*, 119–121.
Westphal, J. (2006). The future and the truth-value links: A common sense view. *Analysis, 66*, 1–9.
Williams, D. C. (1951a). The myth of passage. *The Journal of Philosophy, 48*, 457–472.
Williams, D. C. (1951b). The sea fight tomorrow. In P. Henle, H. M. Kallen, & S. K. Langer (Eds.), *Structure, method and meaning* (pp. 282–306). New York: Liberal Arts Press.
Williamson, T. (2013). *Modal logic as metaphysics*. Oxford: Oxford University Press.
Zimmerman, D. (2008). The privileged present: Defending an "A-theory" of time. In J. Hawthorne, T. Sider, & D. Zimmerman (Eds.), *Contemporary debates in metaphysics* (pp. 211–225). Oxford: Blackwell.
Zimmerman, D. (2011). Presentism and the space-time manifold. In C. Callender (Ed.), *The Oxford handbook of time* (pp. 163–244). Oxford: Oxford University Press.

Index

F. Correia, S. Rosenkranz, *Nothing To Come*, Synthese Library 395,
https://doi.org/10.1007/978-3-319-78704-6

The manufacturer's authorised representative in the EU is Springer Nature Customer Service Centre GmbH, Europaplatz 3, 69115 Heidelberg, Germany. If you have any concerns regarding our products, please contact ProductSafety@springernature.com

Printed and bound by CPI Group (UK) Ltd, Croydon, CR0 4YY
05/10/2026
02228135-0003